Jackgrayce Dutra Nascimento Silva
Marco Valério Jansen Cutrim

Environmental Sustainability Index

Jackgrayce Dutra Nascimento Silva
Marco Valério Jansen Cutrim

Environmental Sustainability Index

A tool to contribute to public development and conservation policies in the Bacanga River Basin, São Luís/MA

ScienciaScripts

Imprint

Cover image: www.ingimage.com

This book is a translation from the original published under ISBN 978-613-9-61926-9.

Publisher:
Sciencia Scripts
is a trademark of
Dodo Books Indian Ocean Ltd. and OmniScriptum S.R.L publishing group

120 High Road, East Finchley, London, N2 9ED, United Kingdom
Str. Armeneasca 28/1, office 1, Chisinau MD-2012, Republic of Moldova, Europe
Printed at: see last page
ISBN: 978-620-7-70738-6

SUMMARY

ACKNOWLEDGMENTS

This work would not have been possible without the participation and collaboration of professionals and friends, who directly or indirectly contributed to its execution. I would particularly like to thank

- To God, for the gift of life, health and peace;
- To Prof. Dr. Marco Valério Jansen Cutrim, for his guidance, patience and valuable information and teachings for my professional and personal life;
- To Prof. Dr. Antônio Carlos Leal de Castro for his contribution and friendship;
- To Prof. Dr. Clàudio Urbano B. Pinheiro for his scientific information;
- To Prof. Dr. Màrcio Vaz, for his precise answers and professionalism;
- To Ronni for his love, understanding, support and hours of relaxation;
- To the friendships built over these two years, Adriana Coimbra and Miqueias Souza;
- The SEMOSP social team, especially Rosângela Rayol, for their understanding and trust;
- To CAEMA, Instituto das Cidades and IBGE;
- To the Maranhão State Research Support Foundation - FAPEMA - for granting the master's degree scholarship.

SUMMARY

ENVIRONMENTAL SUSTAINABILITY INDEX OF WATER USE (ISA) AS A TOOL FOR BUILDING PUBLIC POLICIES IN THE BACANGA RIVER BAY, SÂO LUÌS/MA.

The hydrographic basin is a complex environmental system, resulting from the interrelationships between the physical-natural and socio-economic subsystems. The identification of the watershed as a unifying unit of environmental processes and human interference leads to its adoption as a research unit, allowing for the assessment of environmental quality and sustainability, based on the analysis of physico-chemical, biological and socio-economic factors. The aim of this work is to compose a Water Use Sustainability Index (ISA) which will help to support the adoption of public policies aimed at the sustainable use of water resources. To construct the ISA, two dimensions were analyzed: ecological and socioeconomic. The Ecological Dimension (ED) used physico-chemical indicators (pH, DO, ammonia, nitrate and phosphate), biological indicators (phytoplankton) and the environmental preservation index (IP); the Socioeconomic Dimension (SD) used income, education and sanitation indicators, generating an index for the five sub-basins of the Bacanga river basin and for the basin as a whole. For the physico-chemical and biological indicators, past data from Martins (2008) and Costa (2008) were used. The results of the application of the ISA are based not only on the proposal of a tool for assessing the degree of sustainability of the Bacanga river basin, but also on the ability to point out the main culprits responsible for advances and setbacks in the quest for sustainability. The five sub-basins studied showed significant compromises in terms of sustainability. The lowest index, 0.29, was found for sub-basin 5, revealing weak sustainability. Sub-basins 3 and 4 showed the same value of 0.36, demonstrating that the environmental quality of these environments is compromised. Sub-basins 1 and 2, on the other hand, had a higher index than the others, 0.43 and 0.42 respectively. These sub-basins, despite having an older population and consequently a larger anthropized area, have good services such as water supply and schools. The ISA for the Bacanga river basin was 0.37, showing that the basin is a polluted environment. On the one hand, the basin's capacity to support the physical-natural subsystem is saturated in the areas where it favors urban occupation, a situation caused mainly by high soil sealing and high population concentration; on the other hand, areas of greater natural fragility tend to be occupied by socially excluded people who are more vulnerable to environmental risks, feeding a dynamic in which low quality of life and low environmental quality reinforce each other. The ISA has proved to be a suitable tool for supporting decision-makers and public policymakers in the process of building more sustainable paths for society.

Key words: Sustainability index, indicators, Bacanga river basin.

1 INTRODUCTION

Water resources have been used by man for centuries for various purposes, such as domestic supply, industrial activities, irrigation, electricity production, leisure and recreational activities, as well as serving as a natural recipient of effluents. Population growth and the increase in economic activities around the world have had an impact on aquatic ecosystems and have led to increasingly high levels of pollution.

In Brazil, around 63% of garbage dumps are located near rivers, lakes and sandbanks; sewage and human excrement are important causes of water quality deterioration (MORAES; JORDÂO, 2002).

There are basically two ways in which pollutants reach bodies of water, through point sources and diffuse sources. Point sources reach the aquatic environment in a concentrated form (discharge from a sewage outfall into a river), while diffuse sources spread pollutants over various stretches of the drainage basin (MACÊDO, 2003).

The state of Maranhão has one of the greatest water potentials in the northeast, formed by a series of perennial rivers such as the Mearim, Grajaù, Itapecuru and Pindaré. The state capital, Sâo Luis, has watersheds located in urban areas that are suffering from declining environmental quality, with the Bacanga Watershed being an example of disordered occupation.

The river basin systematically reflects all the effects of actions and degradations resulting from human activities. Thus, the importance of identifying this unit as a unifier of environmental processes and human interference leads to its adoption as a research unit, allowing the assessment of environmental quality and sustainability, based on the analysis of physico-chemical, biological and socio-economic factors (FAY, 2006).

Therefore, studying environmental quality in river basins is fundamental for establishing sustainable development, as it is closely linked to the quality of life of the population.

In order to establish the appropriate analysis parameters for assessing environmental sustainability, it is necessary to develop tools to measure degradation and its capacity for resistance and resilience.

In order to draw up such instruments, all the dimensions involved in the issue of sustainability must be taken into account, such as its ecological, economic and social aspects, so that it is possible to determine the consequences of the various factors involved in the development process on the quality of the environment and people's lives. (BIDONE, 2004)

Studies on development and environmental quality face the challenge of dealing with uncertainty and the lack of systematized information. As a result, many of the decisions made by regulatory bodies in the environmental field are based on imprecise information. In this context, it is necessary to develop work capable of creating reliable indicators that can support crucial and urgent political decision-making.

For this reason, environmental quality indicators can be used not only for comparative evaluation of the quality of life and the environment, but also between cities in the regions that are studied and evaluated. In this way, they can become auxiliary tools in the city planning process, by indicating the areas with the best or worst relative performance, pointing out trends and drawing attention to weak points (AZEVEDO, 2006).

Within this context, it is worth mentioning that several studies have used indicators and created indices to assess environmental sustainability, such as Couto (2007) who generated an environmental sustainability index for river basins in urban areas; Azevedo (2006) who used indicators as a tool for analyzing socio-economic and environmental data to define public policies in the municipality of São Gonçalo, in the state of Rio de Janeiro; Braga *et al.* (2004) described the challenge of measuring sustainability indices for the Piracicaba River Basin (MG). Isla (1998) used urban indicators as a proposal for overcoming the situation in the province of Barcelona/Italy; Ramieri (1998) also used sustainable development indicators for the city and lagoon of Venice and Taylor (1998) used sustainability indicators to implement local Agenda 21 in Venice, Italy.

The work already carried out demonstrates the usefulness of using indicators to assess environmental quality, presenting them as a useful tool for the sustainability and management of a river basin. The study of the Bacanga river basin is therefore of great interest, considering its importance for the island of São Luis, due to its location and its use as a source of drinking water and other resources for the population.

Since 1612, with the arrival of the French in the city, the basin has suffered irregular urban occupation on its banks, as neighborhoods without infrastructure have sprung up over the years. According to MMT (2007), this is the basin on the island of São Luis with the highest population density (80 inhabitants/ha), as well as having the second highest population, second only to the Anil River basin.

In the light of these considerations, a methodology is proposed for generating a water use sustainability index for the Bacanga river basin, aimed at covering a set of factors relating to socio-economic and environmental aspects, with a view to obtaining an index capable of coming as close as possible to the state of environmental quality of its ecosystems and the quality of life to which the populations living in these areas are subject.

Considering the purposes of this work, which has as its basic assumption the construction of environmental assessment mechanisms, the Bacanga river basin was chosen because it is an area that corresponds to an urban watershed and has the characteristics of land use and occupation, socio-economic structure and physical configuration, typical of those found in areas subject to intense urbanization and the need to implement public policies aimed at improving the quality of life of riverside populations.

2 OBJECTIVES

2.1 GENERAL OBJECTIVE

- Use and validate the Environmental Sustainability of Water Use Index (ISA) for the Bacanga Basin for use in public development and conservation policies.

2.2 SPECIFIC OBJECTIVES

- To evaluate the indicators that will make up the Ecological and Socioeconomic Dimensions of the Bacanga river basin;
- Comparative analysis of the results of the application of the sustainability index in five sub-basins of the Bacanga river basin, as a subsidy in the selection of potential conflicts for the adoption of public policies;
- Draw up mitigation measures for the impacts found in the study area.

3 LITERATURE review

Economic growth and environmental preservation are often seen as antagonistic objectives. There is enough evidence to prove that industrialization, the expansion of the agricultural frontier and urbanization create significant pressures on the natural base of an economy, either through the accelerated use of exhaustible natural resources in production processes, or due to the generation of pollution that degrades environmental quality (MOTTA, 1996).

It is also argued, with equally irrefutable evidence, that the nations currently considered the richest have achieved satisfactory levels of growth at the expense of these environmental losses. Therefore, such a growth pattern is inevitable for those nations that are still in the process of development today.

However, the environmental issue should not necessarily be understood within this contradiction. Although it lacks equally strong evidence, there are consolidated theoretical arguments that make it possible to refute the extreme positions mentioned above.

Any productive system depends, directly or indirectly, on natural resources; therefore, respecting environmental limits is no obstacle to development. This realization has led to the perception of environmental risk and the compromising of societies' development possibilities, including the very possibility of economic growth, and therefore the need to implement a new form of development, sustainable development (BIDONE, 2004).

The sustainability of a society's development requires that environmental quantity and quality do not decline over time, thus preventing environmental degradation from jeopardizing future development possibilities.

According to the WECD (1987), sustainable development is development that meets the needs of the present without compromising the ability of future generations to meet their own needs.

In this sense, there are two currents: one that follows the pattern of "weak sustainability", in which we find the hypothesis that there is perfect substitutability between natural capital and material capital. In other words, there is the technological possibility of continued growth, as long as part of the economic income is reinvested in order to maintain the total level of capital (natural and material) in an economy. In this way, future generations are guaranteed a productive capacity (and access to natural resources for direct consumption) equivalent to what is available to the present generation. (AZEVEDO, 2004)

At the other extreme is the "strong sustainability" pattern, in which the two forms of capital are not substitutable and, therefore, sustainable growth would only occur if the level of the natural capital stock were kept constant. In other words, since the technological possibilities of compensating for losses of natural capital with material capital would be very limited, the well-being of future generations would only be guaranteed if the stock of natural capital were kept intact. (MOTTA, 1996)

This sustainability would be constantly measured by analyzing the dimensions involved in the process, such as its ecological and socio-economic aspects through Environmental Sustainability Indices (ISA).

In order to build the ISA, it is necessary to use indicators that will direct the comparative assessment of the quality of life and the environment, constituting an auxiliary tool in the planning process by indicating the areas with the best or worst relative performance, pointing out trends and drawing attention to weak points. (COUTO, 2007)

3.1 CONSTRUCTION OF ENVIRONMENTAL INDICATORS

At the end of the 1980s, the first proposals for the construction of environmental indicators emerged. These proposals had the common objective of providing input for the formulation of national policies and international agreements, as well as for decision-making by public and private actors. In addition, they sought to describe the interaction between human activity and the environment, as well as giving the concept of sustainability greater concreteness and functionality (BRAGA *et al.*, 2004).

Attempts to construct environmental and sustainability indicators have been divided into two strands: the biocentric strand, which consists of the search for biological, physical and chemical indicators of the ecological balance of ecosystems; the second, the economic strand, which consists of monetary evaluations of natural capital and the use of natural resources; and the third strand, which seeks to construct indicators of sustainability and environmental quality that combine aspects of the natural ecosystem with aspects of the economic system and the quality of human life. (ESI, 2002)

Third environmental indicators are, as a rule, models of anthropogenic activity/environment interaction that can be classified into three main types: state, pressure and response. While state indicators seek to describe the present physical or biological situation of natural systems, pressure indicators attempt to measure/evaluate the pressures exerted by anthropogenic activities on natural systems, and so-called response indicators seek to assess the quality of policies and agreements formulated to respond to anthropogenic impacts and minimize them (HERCULANO, 1998).

In general, their evaluations focus on the short/medium term and face methodological difficulties in terms of conceptualization, defining variables and obtaining and processing data.

A first difficulty concerns the different concepts and conceptions of what sustainability and environmental quality are, which obscures the process of choosing the variables to be used in measurement. Isla (1998) points out that in the case of local environmental indicators, due to the lack of a conceptual definition, what we see are lists of isolated indicators without clear causal and hierarchical relationships, rather than a coherent system for measuring/evaluating the phenomenon. Therefore, a crucial first step in the construction of indicators and indices is to make explicit the conceptual approach used.

3.2 USES OF ENVIRONMENTAL INDICATORS

Studies on environmental quality and development often face the challenge of dealing with uncertainty and the lack of systematized information. Many of the decisions made by regulatory bodies in the environmental field are based on imprecise information and fragile certainties (ROSS, 1994).

The nature of the relationship between the environment and development is the subject of controversy and uncertainty. In this context, empirical work capable of creating reliable indicators that can be used as a basis for studies and political decision-making is crucial and urgent.

The concept of sustainability, or sustainable development, although widely used over the last two decades to the point of becoming an obligatory reference in academic, political and cultural debates, is far from having a consensual meaning. It is more a concept in permanent construction and reconstruction, a symbolic battlefield and a powerful *marketing* tool than a consolidated reference of standards for the relationship between environmental conservation and economic growth. (COUTO, 2007)

What underlies the various versions of what sustainable development would be is an approximation of the environmental, social and economic dimensions of the environment.

A city or place is considered to be more or less sustainable to the extent that it is able to maintain or improve the health of its environmental system, mitigate degradation and anthropogenic impact, reduce social inequality and provide its inhabitants with basic living conditions, as well as a healthy and safe built environment, and to build political pacts that allow it to face present and future challenges. (ROSS, 1994)

The environmental issue can be dealt with through sustainable development indicators that should highlight the ownership of the local ecosystem and not ignore the social and economic facts pertinent to human development. These indicators should subsidize strategic options when defining policies and act as instruments to support governance in the region. (ISAIAS, 2008)

The concept of sustainable development, as set out in the Brundtland Report and Agenda 21, is presented from a multidimensional perspective, combining social, economic, environmental and political elements.

Identifying, constructing and validating sustainability indicators for a given location makes it possible to see, for example, where public policies and initiatives by society have been successful or not and, above all, to assess the path towards more sustainable development. The support provided by the use of sustainability indicators will make it possible to correct course, decentralize more effectively, locate the main shortcomings and address them efficiently (BIDONE, 2004).

According to Benetti (2006), each indicator has advantages and disadvantages, but the most important thing is that indicators are an important parameter for guiding management and planning policies and actions that can be developed to deepen commitment to the targets set. The IBGE (2002) defines indicators as tools made up of one or more variables which, when associated in various ways, reveal broader meanings about the phenomena they refer to.

The basic task of an indicator is to express, in the simplest possible way, a particular situation that you want to evaluate, and the result of an indicator is a snapshot of a particular moment, and shows, on a measured basis, what is being done, or what you want to be done. (FERNANDES, 2004)

The use of indicators to assess the dynamics of a complex system (environment, organization, territory, etc.) must take into account the essential objectives for which they were designed (TURNES, 2004). However, it should be noted that there is no type of measure that can accurately describe all the aspects, structure and dynamics of issues related to the environment.

The use of indicators and indices has been the subject of controversy in technical/scientific forums, due to the simplifications that are made in the application of these methodologies. The choice or use of indicators often requires differentiating their levels of importance or their weighting in order to signal their degree of priority for the established objectives (MAGALHÂES - JÛNIOR *et al.*, 2003).

Turnes (2004) highlights some of the advantages and limitations of applying indicators. The advantages include: assessing levels of sustainable development; the ability to synthesize information of a technical/scientific nature; identifying key variables in the system; ease of transmitting information; use as a tool to support decision-making and environmental management processes; the possibility of identifying trends; and comparison with pre-defined standards or targets.

Limitations include: lack of basic information; difficulty in defining mathematical expressions that best translate the selected parameters, loss of vital information in the data aggregation processes; measurement of what is measurable, rather than what is actually measurable; correct interpretation of the measures; different criteria for defining the limits of variation of the index in relation to the established impositions; lack of robust criteria for selecting some indicators; human, financial and time limits; and difficulty of application in certain areas, such as land use planning and landscape.

Fidalgo (2003) suggests some characteristics that indicators should have: accessibility of data; clarity in communication; relevance, i.e. they must reflect something basic and fundamental; geographical scope; standardization, to increase the possibility of comparing one reality with others; predictivity, to enable the decision-maker to know problems in advance; measurable, they must be easily measurable and capable of being monitored regularly at a feasible cost; and consistent, well supported in technical and scientific terms.

4 MATERIAL AND METHODS

4.1 Characterization of the study area

Maranhão is located in the transition area between equatorial and tropical climates. The city of SSo Luis, the state capital, has a tropical climate of type AW, according to the classification proposed by Koppen, with average temperatures between 19°C and 28°C, average rainfall of just under 2,000 mm/year and two well-defined seasons: rainy from January to June and dry from July to December. The average annual relative humidity is between 75 and 90% (MMT, 2007; RHAMA, 2008).

The city of SSo Luis is located on the island of SSo Luis, together with the municipalities of Raposa, SSo José de Ribamar and Paço do Lumiar. According to the IBGE (2008), the city of SSo Luis has a territorial area of 827 km^2 and a population of 957,515 inhabitants, making it the main city in the Greater SSo Luis Metropolitan Region.

The Bacanga river basin corresponds to an integrated and interdependent biophysical and socio-economic system, including agricultural, industrial and service activities, plant formations, springs, among others (OLIVEIRA, 2008). It covers an area of approximately 110 km2, occupying the northwestern portion of the island of SSo Luis - MA, with coordinates 2° 32' 26" - 2° 38' 07" S and 44° 16' 00" - 44° 19' 16"W (MMT, 2007).

It is bordered to the north by the SSo Marcos bay and the Anil river basin; to the south by the central plateau of the island in the Tiririca1 region; to the east by the watershed separating the Anil, Paciência and Cachorros river basins; and to the west by the Itaqui basin (COELHO, 2006) (Figure 4.1).

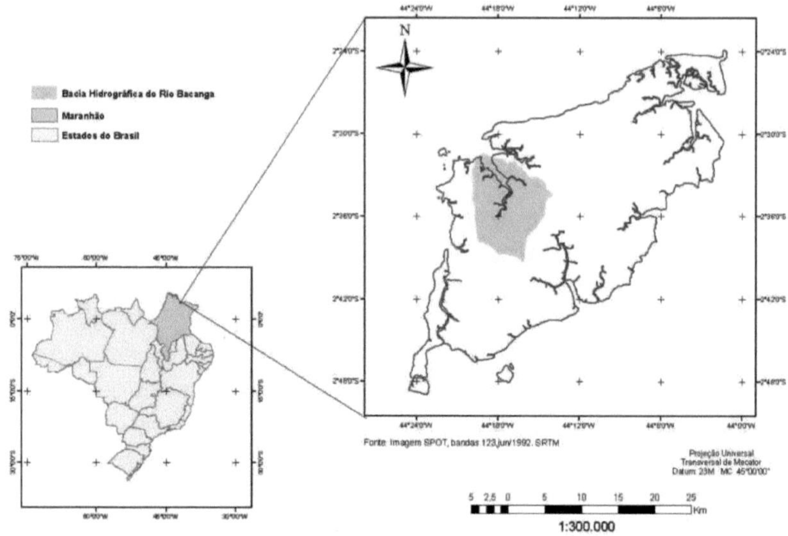

Source: OLIVEIRA, 2008

FIGURA 4.1 **Location of the Bacanga watershed, Sao Luis Island/MA.**

The Bacanga River is the main hydrological component of the Bacanga River basin, with a hydrological regime dependent on seasonal rainfall and tidal levels, where the participation of fresh water is minimal (LABOHIDRO, 1998). It has its diffuse sources in neighborhoods close to the Maracanâ region, covering a distance of approximately 22 km. The Gapara and Bicas rivers are some of its most important tributaries.

The Bacanga river basin is the second most heavily urbanized basin in the city of Sâo Luis, second only to the Anil river basin, and presents problems associated with disorderly occupation and sewage disposal. Despite informal urban occupation, the basin has an extensive area of vegetation cover, forming the Bacanga State Park, where the Batatâ dam is located, which produces 10% of Sâo Luis' drinking water, and the Maracanâ Environmental Protection Area, created as a buffer area for urban and industrial expansion.

In addition, the area is home to the city's historic center, the main industrial region, as well as a high percentage of the low-income population.

The demographic concentration of the Bacanga river basin in 2000 was 195,353 inhabitants, just over 20% of the total population of the municipality of São Luis in the same year (MTT, 2007). According to the IBGE (2008), the population of the island of São Luis has an annual growth rate of 2%, so it is estimated that the population for the basin area in 2009 is around 299,945 inhabitants, irregularly distributed.

In the study, the Bacanga river basin was subdivided into 5 sub-basins (Figure 4.2) according to the main water flows and neighborhoods, as described below:

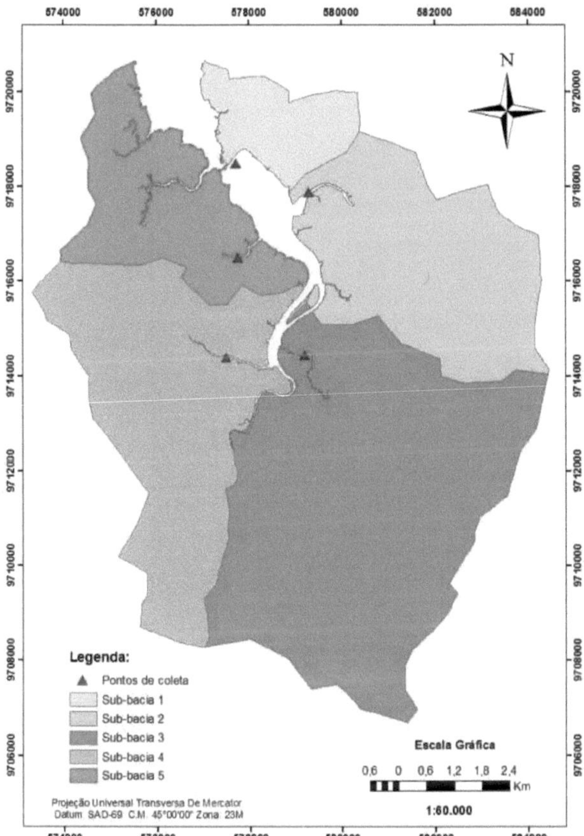

FIGURA 4.2 - **Location of the 5 sub-basins of the Bacanga river basin, Sao Luis/MA and collection points.**

4.1.1 Sub-basin 1

It corresponds to one of the most urbanized regions of the basin, where the historic center and some of the oldest neighbourhoods in São Luis are located. Many houses were built on the banks of the river, below the 4m level, and during the rainy season when the Bacanga dam floodgates are not opened, these houses are flooded. You can see the mouth of the Bacanga, and mangrove vegetation on the banks (Figure 4.3).

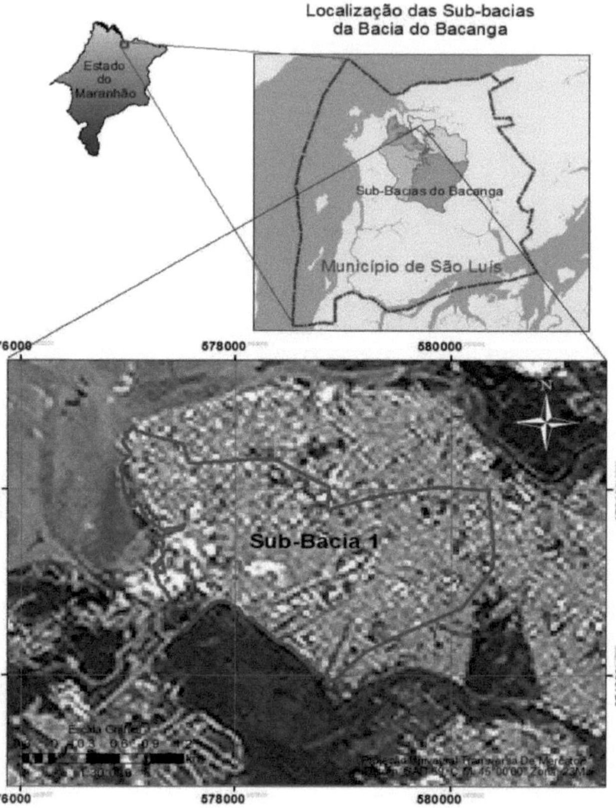

FIGURE 4.3 - **Location of sub-basin 1, Bacanga river basin, Sao Luis/MA.**

4.1.2 Sub-basin 2

The area most exposed to environmental degradation is where most of the city's sewage is discharged. There are some old neighborhoods, such as Coheb, Fâtima, Filipinho, and Sitio do Fisico. The vegetation cover is characterized by low vegetation, as well as areas of crops. The main tributaries are the Rio Bicas and the Igarapé Coelho (Figure 4.4).

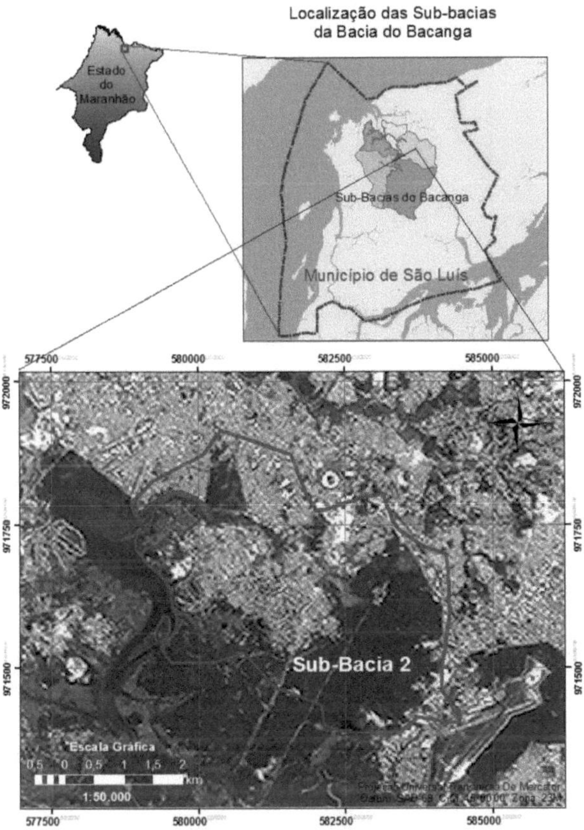

FIGURE 4.4 - **Location of sub-basin 2, Bacanga river basin, Sao Luis/MA.**

4.1.3 Sub-basin 3

Located near the Bacanga State Park, it is characterized by having few dwellings, although there are the Vila Itamar and Recanto Verde housing estates. This sub-basin includes the Batatã stream and the Mamão Igarapé (Figure 4.5).

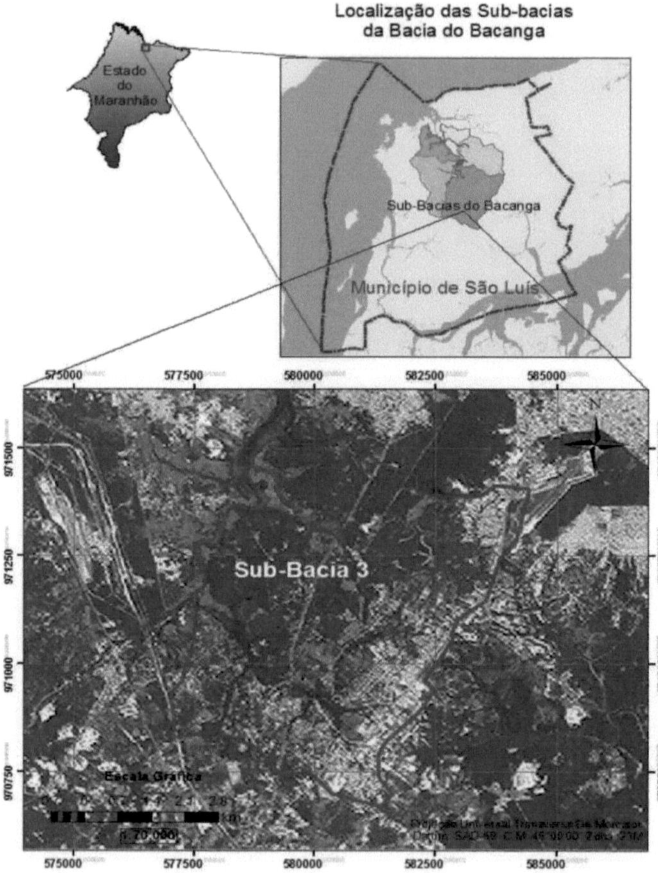

FIGURE 4.5 - **Location of sub-basin 3, Bacanga river basin, Sao Luis/MA.**

4.1.4 Sub-basin 4

There are few houses on its banks, but some places are deforested as a result of rice cultivation, which is planted by the community near the mouth of the Gapara River. This sub-basin is home to a very populous neighborhood, Vila Embratel (Figure 4.6).

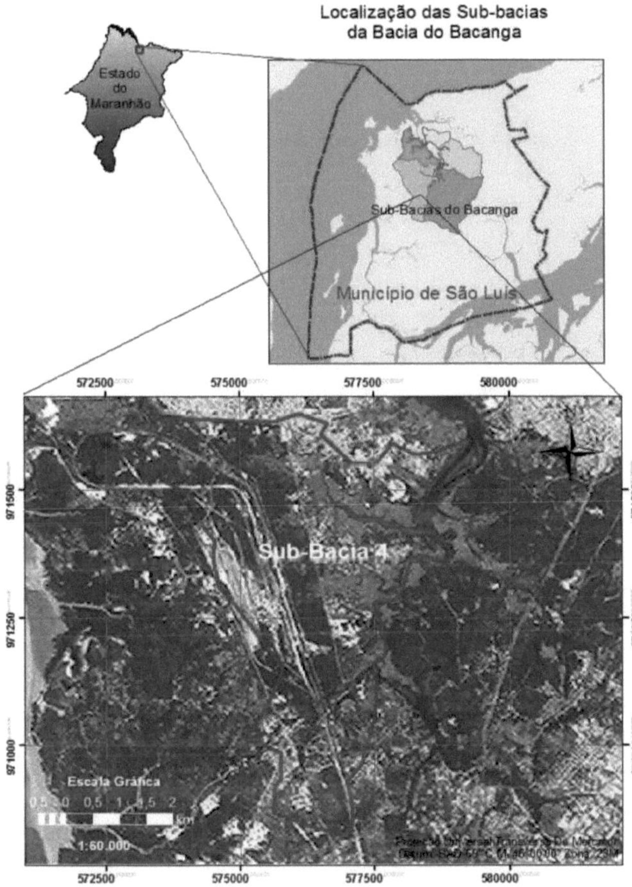

FIGURE 4.6 - **Location of sub-basin 4, Bacanga river basin, Sao Luis/MA.**

4.1.5 Sub-basin 5

There is a strong urban influence, with many houses located on the banks of the watercourse, which have increasingly replaced the mangrove vegetation. The area is home to some newer but highly populated neighborhoods, such as Anjo da Guarda, Sà Viana, Vila Nova and the UFMA University Campus (Figure 4.7).

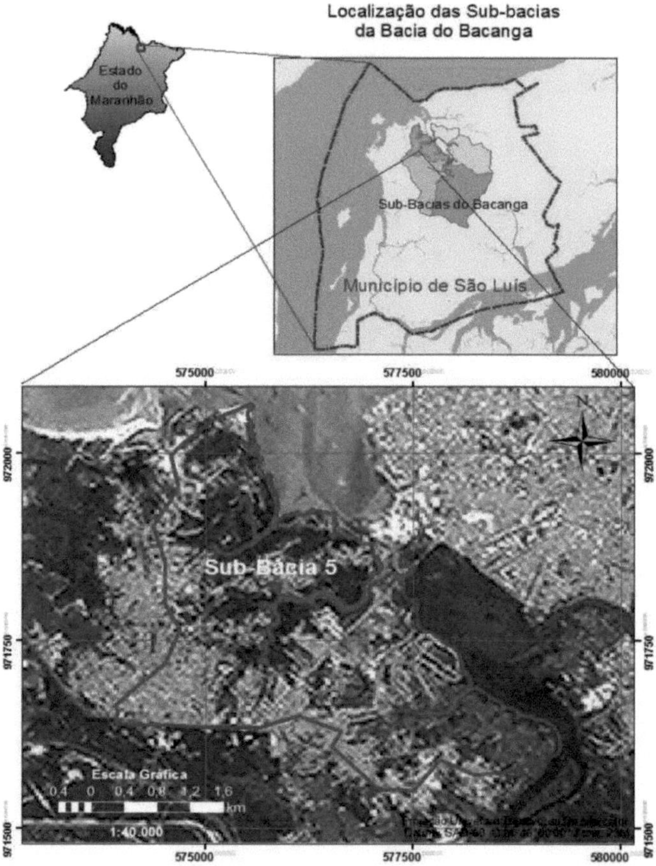

FIGURE 4.7 - **Location of sub-basin 5, Bacanga river basin, Sao Luis/MA.**

4.2 Methodology

4.2.1 Interpreting past data

This part of the work is based on the interpretation of previous data from Martins (2008) and Costa (2008) on the physico-chemical analysis of water and phytoplankton composition. These authors carried out bimonthly samples (October 2007 to June 2008) at fixed, pre-determined points along the Bacanga river estuary (Appendix A).

The collection points were distributed along the five sub-basins represented by: Barragem (sub-basin 1), Rio das Bicas (sub-basin 2), Igarapé do Mamâo (sub-basin 3), Rio Gapara (sub-basin 4), Igarapé Jambeiro (sub-basin 5).

4.2.2 Construction of the Water Use Environmental Sustainability Index (ISA)

The methodological proposal of the Water Use Sustainability Index (ISA) focuses on assessing the environmental sustainability of water resources by sub-basin, using past and current data, creating measurement tools such as environmental sustainability indicators.

The ISA is made up of two dimensions, ecological and socio-economic. To calculate the index, we used the methodologies of Isaias (2008) and Fay (2006) with modifications, as well as census data provided by the Institute of Cities and IBGE. Each dimension has indicators that will reflect the reality of each of the five sub-basins, to finally make up the ISA of the Bacanga basin.

The same weight was given to the dimensions involved in the index when calculating the ISA for each sub-basin of the Bacanga basin. Each indicator represents the dimensions involved in the formula and ranges from 0 to 1, where 0 means the worst situation of the indicator and 1 the best, with the final result of each indicator being the average of its values, according to the methodology proposed by Isaias (2008).

This gives us the following formula (1):

$$ISA = \frac{\{DE\ (0\text{-}1) + DS\ (0\text{-}1)\}}{2}$$

Where:

ISA = Environmental Sustainability Index for Water Use

DE = Ecological Dimension

SD = Socio-economic Dimension

The composition and calculation of the ISA will be presented in the following sections. Table 01 summarizes the dimensions, indicators and variables that make up the ISAsub and ISA.

TABLE 4.1- **Summary of the ISA calculation by sub-basin and the ISA for the Bacanga river basin.**

CONTENTS	DIMENSIONS	INDICATORS AND VARIABLES
ISAsub = **{DE (0-1) + DS (0-1)} 2**	Ecological (DE) $DE = \frac{\{VF\ (0\text{-}1) + VB\ (0\text{-}1) + IP\ (0\text{-}1)\}}{3}$	DE = Ecological Dimension VF = Physical and chemical water variables VB = Variables Biological IP =

		Environmental Preservation Index
	Socio-economic (SD) DS = {IR (0-1) + IE (0-1) + IS (0-1)} 3	SD = Socio-economic dimension; IR = Income indicator; IE = Education indicator; IS = Sanitation indicator

ISA = (ISAsubl + ISAsub2 + ISAsub3 + ISAsub4 + ISAsub5) 5

Source: Adapted from ISAIAS (2008).

The Water Use Sustainability Index (ISA) for the basin as a whole is the average of all the ISAsub, considering the total number of sub-basins. For this purpose, we have the following formula (2):

$$ISA = \frac{(ISAsub1 + ISAsub2 + ISAsub3 + ISAsub4 + ISAsub5)}{5}$$

Where:

ISA = Water Sustainability Index

ISAsub = Each sub-basin that makes up the unit under study.

In this way, the Water Use Sustainability Index (WSAI) can be calculated for each sub-basin and for the basin as a whole. The ISA of the sub-basins allows the most critical areas to be identified from the point of view of sustainability within the basin.

4.2.3 Ecological Dimension (ED)

Water quality standards are the physical, chemical and biological characteristics that are desirable according to the preponderant uses established by rules defined by society. Preponderant uses are the beneficial uses determined for a certain body of water, *while* beneficial uses are those that promote economic benefits and/or the well-being and good health of the population.

The ecological dimension analyzed the environmental quality of natural resources, applying indicators simultaneously in order to obtain an integrated picture of the system, since different indicators focus on different aspects of ecosystem integrity. This dimension is constructed through the integrated analysis of 03 indicators, represented by:

4.2.3.1 Physico-chemical water variables (VF)

For the physico-chemical variables, past data from Martins (2008) was used to later classify them in terms of environmental quality, as the best condition (1), those with values equal to those established by CONAMA Resolution 357/05, or the worst condition (0), those in disagreement with the Resolution.

The classification adopted from CONAMA Resolution 357/05 (Table 02) refers to brackish waters that belong to class 1, intended for:

- primary contact recreation;
- protection of aquatic communities;
- aquaculture and fishing;
- irrigation of vegetables that are eaten raw and fruit that grows close to the ground and is eaten raw without removing the skin;
- irrigation of parks, gardens, sports and leisure grounds with which the public may come into direct contact.

TABLE 4.2 - **Physico-chemical variables and values recommended by Resolution CONAMA 357/05.**

VARIABLES	pH	OD	Nitrate	Ammonia
CONAMA 357/05	6,5-8,5	≥ 5 mg/L	≤ 0.4 mg/L	≤ 0.4 mg/L

For the classification of phosphate, the methodology proposed by Baumgarten *et al.* (1996) was used, whose value adopted for unpolluted estuaries is 1.2μM, this value being the best condition (1) and different values (0).

The indicators used were: hydrogenionic potential (pH), dissolved oxygen and nutrients (ammonia, nitrate and phosphate).

- Hydrogen potential

The hydrogen potential of the water was determined *in situ* using a Hanna digital ph meter.

- Dissolved oxygen in water

The dissolved oxygen concentration (mg/L) was determined according to the Winkler method (modified by GOLTERMAN *et al.*, 1978).

- Nutrient salts

Nutrient salt analyses were carried out in the laboratory using the following methodologies: Ammonium - Indufenol blue method described by Koroleff (1976) and cited in Aminot; Chaussepied (1982); nitrate, method of Wood *et al.* (1967), cited by Aminot; Chaussepied (1983); phosphate, method of Murphy; Riley (1962) through the phosphate paramobilidate reaction.

4.2.3.2 Biological variables: phytoplankton density (VB)

The biological variables used were phytoplankton density, which corresponds to the number of cells/liter.

- Phytoplankton density:

Counting was carried out under a Zeiss inverted microscope, at 400x magnification, according to the Utermohl method (1958), using the technique of counting by illuminated fields, systematized in a cross. The counting results were used in the formula (3) below:

Ncub = (Nind / Ncamp) - (A2 / a2), where:

Ncub = Total number of individuals observed in the cuvette;

Nind = Number of individuals observed in the random fields;

Ncamp = Number of illuminated fields counted (100 to 120 fields);

A = Area of the cuvette;

a = Area of the illuminated field at 400x magnification; (A2 / a2) = 2,406.62.

The number of individuals per sample volume was estimated using a simple rule of three and the results are expressed in cells.L^{-1}

From the cells/liter values, the following indices were calculated:

- Diversity:

Shannon's index (1948) was used to calculate diversity (H'), expressed by the following formula (4):

$$H' = - \Sigma \, Pi \bullet \ln Pi \qquad Pi = Ni / N, \text{ onde}$$

Pi = Probability of collecting species i in the population;

Ni = Number of individuals of each species;

N = Total number of individuals;

H' = Shannon's diversity index, in bits.cel^{-1} (1 nits.cel^{-1} = 0.693 bits.cel).$^{-1}$

The values (MARGALEF, 1958):

≥ 2.1nits.cel^{-1} - high diversity = 1

≥ 1.4 to < 2.1 nits.cel^{-1} - average diversity = 0.7

≥ 0.7 to < 1.4 nits.cel^{-1} - low diversity = 0.3

< 0.7 nits.cel^{-1} - very low diversity = 0

4.2.3.3 Environmental Preservation Index (IP) (BITTON, 2000 - Modified)

Land cover indicates the level of degradation of the drainage basin, showing the classes that exist in these places, such as urbanized areas, cultivation activities, secondary forests and waterlogged areas. A basin with more anthropogenic activities and less vegetation cover would be more environmentally compromised.

The river drainage of a basin is made up of a set of interrelated flow channels that form the drainage basin. The amount of water that reaches the river courses depends on the area occupied by the basin, the total rainfall and its regime and the losses due to evapotranspiration and infiltration (ALCÂNTARA, 2004).

The contour of the basins is not a rigid boundary, depending on the flow and the type of soil. Underground runoff often alters the surface definition of the watershed, particularly in highly permeable soils or soils with abundant cavities, as is the case in karst and volcanic regions.

Morphological characteristics of river basins that influence the runoff process: the shape of the basin, which is very important because of its influence on flooding. Relief is another factor that largely influences the shape of runoff (LANA *et al.*, 2001).

Drainage density (Dd) is the ratio between the total length of watercourses in a hydrographic basin and its area. Estimating drainage density depends on the scale of the topographic map you are working with, as smaller scale maps do not show the watercourses of elementary basins.

Depending on whether the drainage density is high or low, it will generally represent an impermeable watershed with very steep terrain or a very permeable watershed with gentle terrain, respectively. Drainage density varies directly with the extent of surface runoff and thus provides an indication of the efficiency of the basin's natural drainage.

- Dd = 0.5 km/km^2 Poorly drained basin
- Dd = 3.5 km/km^2 Well-drained basin

Basins with a higher drainage density tend to be more prone to flooding than basins with a lower drainage density.

The calculation of this index will analyze the conservation status of the sub-basins and the basin as a whole, represented by the greater occurrence of natural cover. A basin with a drainage density of 1 and a natural cover 10 times greater than that altered by man is a well-preserved drainage basin (IP = 10.0).

Drainage density is represented by the ratio between the total length of river courses in a drainage basin and the total area of the same basin. According to formula (5):

$$Dd = \frac{CR}{AT}$$

Where:

Dd = Drainage density

CR = River courses in a drainage basin

AT = Total area of the basin

The Environmental Preservation Index (IP) is represented by the following formula (6):

$$IP = \frac{(1)}{Dd} \times \left[\frac{(AV + AE)}{(AA)}\right]$$

Where:

Dd = Drainage density

AV = Vegetated Area

AE = Percentage of waterlogged areas

AA = Percentage of anthropized areas

According to the work proposal, the following descriptions are considered to be the main characteristics of each class:

- Secondary Forest: representing areas of shrub and tree vegetation.
- Anthropized areas: these are densely populated urban areas, residential areas with neighbourhoods (villas or invasions) with little or no infrastructure, no sewage system, with or without asphalt, and areas with great potential for surface flow with high impermeability, as well as exposed soil and agricultural areas.
- Waterlogged areas: these represent areas where the upper limit of the water table is close to the land surface. The vegetation was characterized as riparian forest and mangroves.

The calculation of the ecological dimension is represented by the formula (7):

$$DE = \frac{\{VF\ (0\text{-}1) + VB\ (0\text{-}1) + IP\ (0\text{-}1)\}}{3}$$

Where:

DE = Ecological Dimension

VF = Physico-chemical water variables

VB = Biological variables

IP = Environmental Preservation Index

4.2.4 Socio-economic Dimension (SD)

It expresses the social and economic differences between the sub-basins; 03 indicators were used for this profile. To do this, we used past data from the São Luis City Hall and secondary data from the IBGE, from the neighborhoods that make up the basin (Annex B), which represented the demographic characteristics of the community, portraying their desires and the provision of basic services. The data referring to this dimension is from the year 2000, as the census only took place in 2010.

4.2.4.1 Income Indicator (IR)

The income indicator was based on per capita income data for the neighborhoods that make up each sub-basin, provided by the IBGE (2000) and the Institute of Cities. This indicator will allow an analysis of the population in terms of income and will indirectly reflect the population's environmental awareness. For this purpose, the following scores will be used, as shown in table 4.3.

TABLE 4.3 - **Household income indicator values and their respective scores.**

INCOME INDICATOR (minimum wage)			
$0 \leq IR < 1$ = Value 0	$1 \leq IR \leq 3$ = Value 0.25	$3 < IR \leq 5$ = Value 0.75	$IR > 5$ = Value 1

4.2.4.2 Education Indicator (IE)

The education indicator was based on data provided by IBGE (2000) and Instituto das Cidades for each neighborhood that makes up the sub-basins (Annex C). This indicator is scored on four variables (Table 4.4) which will show the level of education of the population and indirectly the environmental awareness of the population.

TABLE 4.4. **Education Indicator values and their respective scores.**

EDUCATION INDICATOR/IE			
No instruction = 0	Primary Education = 0.3	Secondary Education = 0.7	Higher education = 1

4.2.4.3 Sanitation Indicator (SI)

The sanitation indicator includes urban services made available to the area's population, such as water supply and sewage systems. This indicator will show whether these essential services are being promoted by the public authorities.

The calculation used data on water supply and sewage provided by the Companhia de Agua e Esgoto do Maranhâo (CAEMA), which corresponds to the percentage of these services provided to the neighborhoods in each sub-basin, with the indicator being the average of these services per sub-basin.

4.2.5 Making thematic maps

Thematic maps of the sub-basins and the Bacanga river basin were used to create the maps:

- Landsat-5, TM, multispectral images, bands 1, 2, 4 and 7, referring to 20/06/1984 and 10/09/2008, orbit/point: 220/62;
- Digital file, in shapefile format, referring to the theme: political territorial division, produced by the Brazilian Institute of Geography and Statistics - IBGE (2006);
- Digital file, in shapefile format, referring to the theme: Conservation Units, produced by the State Secretariat for the Environment and Natural Resources of Maranhão - SEMA
- Image processing program, ER MAPPER 6.4;
- Image processing program, SPRING 5.1.3;
- Program for producing thematic maps, ArcGis 9.2.

For the digital image processing in this work, the TM sensor of the Landsat 5 satellite was used. This sensor has a spatial resolution of 30 m, so it allows us, on a regional scale, to visualize different classes of land cover, use and occupation. It also has a spectral resolution of 7 bands, so it is possible to differentiate between the different spectral signatures of the areas under study.

The scale used was 1:100,000 m for the visualization map. However, the vectorization of the land use units was done at a resolution of 30 m.

The images were first acquired from the INPE website and registered in ER MAPPER 6.4. A database was then created in the SPRING 5.1.3 program, with these specifications: UTM projection system; Earth models in DATUM - SAD69; zone 23M; southern hemisphere. With the respective plane coordinates: 537050/773540 East and 9579010/9780850 North.

In SPRING 5.1.3, some restoration, filtering and contrast functions were performed on the images under study in order to improve and highlight the targets in question. This resulted in Supervised classification using the maximum likelihood technique (MAXVER).

The shapefiles created in the previous stage were imported into the ArcGis 9.2 program, making it possible to create the maps in the program's LAYOUT VIEW function.

5 Results and discussion

5.1 Sub-basin 1

Sub-basin 1 is located close to the Bacanga dam area (mouth of the river). Occupation of this sub-basin began around 1612 with the French invasion. It is home to some of the oldest neighborhoods in the city of Sâo Luis, such as Centro, Madre Deus, Desterro and Bairro de Fàtima, which were first settled on the banks of the Bacanga river basin.

5.1.1 Ecological Dimension (ED)

5.1.1.1 Physico-chemical variables (PV)

Environmental variables are of great importance in complementing the understanding of the situation of a given water body, revealing characteristics of its watershed.

Among the physico-chemical variables, pH was the only one that obtained a satisfactory value for this sub-basin, since it is in accordance with CONAMA Resolution 357/05, and is very useful as it provides us with information about the quality of the water in the environment (Table 5.1).

TABLE 5.1 - **Values of the physico-chemical variables for sub-basin 1, average value of the samples, CONAMA 357/05 reference value, indicator value for the variable and average of the indicators.**

PHYSICAL-CHEMICAL VARIABLES - VF			
VARIABLES	**AVERAGE VALUE X± SD**	**CONAMA 357/05**	**VALUE OF INDICATOR**
pH	7,96 (±0,59)	6,5 - 8,5	1
OD	4.16 (±1.53) mg/L	≥ 5 mg/L	0
AMONIA	0.77 (±0.67) mg/L	≤ 0.4 mg/L	0
NITRATE	0.99 (± 1.40) mg/L	≤ 0.4 mg/L	0
PHOSPHATE	2.69 (±3.46) μM	1.2 μM	0
AVERAGE INDICATORS			**0,20**

Very acidic or very alkaline pH measurements are associated with industrial and domestic discharges. In addition to physical (water transparency) and chemical (different precipitated compounds) changes, the acidification of water bodies causes serious changes in the structure of aquatic communities (ESTEVES, 1998).

Dissolved oxygen is a sensitive indicator of biological and chemical phenomena and is one of the most important variables in environmental characterization (HORNE; GOLDMAN, 1994).

The main sources of oxygen for the water are the atmosphere and photosynthesis, while the losses are consumption through decomposition of organic matter, losses to the atmosphere, respiration of aquatic organisms and oxidation of metal ions. The DO showed a low result (4.16), revealing that there may be intense bacterial activity decomposing the organic matter discharged into the body of water, given that sub-basin 1 is a highly urbanized area (Table 5.1).

The values for the nutrients ammonia (0.77 mg/L), nitrate (0.99 mg/L) and phosphate (2.6 μM) were high in relation to CONAMA Resolution 357/2005 (Table 5.1). Although ammonia is a toxic, non-persistent and non-cumulative substance, in high concentrations it can cause fish to suffocate (MARTINS, 2008).

The presence of this substance indicates recent pollution, since ammonia is the first stage to reach nitrate. In polluted areas, its content is usually high (ESTEVES, 1998).

According to LABOHIDRO (1998), the main supplier of ammoniacal nitrogen and other nutrients to the Sâo Luis estuarine system is the city's domestic sewage.

5.1.1.2 Biological Variable (BV)

According to MARGALEF (1983) coastal regions usually have a low specific diversity of 0.7 to 1.7 nits.cel $^{\wedge 1}$, which can be even lower in estuaries, in early successional communities and in polluted environments. The Bacanga dam region is the closest point to the mouth of the Bacanga river basin and suffers the greatest influence from salinity.

The value of the biological variable observed in sub-basin 1 was 0.5, considered an average value, indicating that this environment is in a process of degradation (Table 5.2).

Costa (2008) carried out a study in the Bacanga river basin, obtaining the highest diversity value (H') of 2.45 bits.cel^{-1} in the Bicas tributary and the lowest value of 0.59 bits.cel^{-1} in the Jambeiro tributary. The same study showed a predominance of diatoms, which are eurialine organisms.

TABLE 5.2 - **Values of the biological variable per collection (density - Shanonn Diversity Index), value of the indicator per collection and average of the indicators for sub-basin 1.**

BIOLOGICAL VARIABLE - PHYTOPLANKTON		
COLLECTIONS	DIVERSITY INDEX DE SHANNON (nits.cel)$^{-1}$	VALUE OF INDICATOR

1	1,71	0,7
2	1,66	0,7
3	1,32	0,3
4	1,03	0,3
INDICATOR AVERAGE		**0,50**

According to Tundisi (1970), the agitation of the waters by the winds causes seasonal changes in the transparency of estuarine water and influences the biodiversity, abundance and primary production of phytoplankton, thus interacting strongly with benthic and planktonic communities in shallow areas, controlling the composition of phytoplankton (BERGESCH; ODEBRECHT, 1997).

5.1.1.3 Environmental Preservation Index (IP)

Sub-basin 1 has an area of 5.29 km^2 ; the total length of the river courses is 2.98 km, with a drainage density (Dd) of around 0.56 km/km^2 , considered low (Table 5.3). Areas of low drainage density ($\leq 0.5\ km/km^2$) are characterized by flat, gentle terrain, whose high permeability allows for rapid water infiltration and the consequent formation of aquifers (VILLELA; MATOS, 1975).

Disorderly population growth over the years in sub-basin 1 has led to the occupation of the lower areas, where the mangroves and wetlands are located.

The rainfall regime shows little surface runoff, generating water erosion mechanisms linked to the initial process of the raindrop and causing laminar or sheet erosion, resulting from the friction of the surface runoff itself, which carries eroded material from points below the slopes to the river channels (BRANCO, 1986).

TABLE 5.3 - **Environmental Preservation Index (IP) values for sub-basin 1.**

ENVIRONMENTAL PRESERVATION INDEX - IP	
LAND USE CLASSES	**VALUES**
Drainage Density (Dd)	0.56 Km/Km^2
Vegetated Area (VA)	0.002 Km^2
Soaked Areas (AE)	0.10 Km^2
Anthropized Areas (AA)	5.18 Km^2
Environmental Preservation Index (IP)	**0.03** Km^2

For the IP, a result of 0.03 Km was obtained^2 , which indicates a high level of degradation for this sub-basin (Table 5.3). Since this basin has a much higher anthropized area value than the vegetated area or flooded areas, i.e. a basin with a higher drainage density, greater anthropogenic occupation and less forest cover, it would be more environmentally compromised (Figure 5.1).

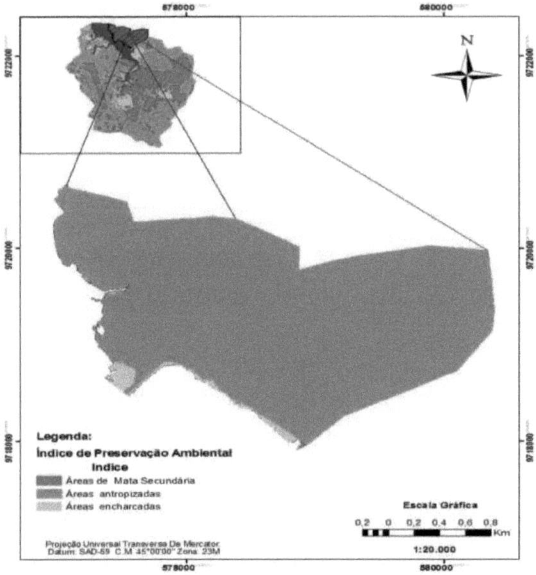

FIGURE 5.1- **Graphical representation of the Environmental Preservation index (IP) for sub-basin 1.**

For the ecological dimension (ED), an index of 0.24 was obtained, showing poor sustainability. This is due to the fact that this sub-basin has very old population centers, dating back to the time of the colonization of São Luis, thus causing damage to the vegetation and water quality.

5.1.2 Socio-economic Dimension (SD)

5.1.2.1 Income Indicator (IR)

The result of the income indicator, 0.48, showed that sub-basin 1 is made up of a population with an average income level, although some neighborhoods such as Desterro, Vila Passos and Bom Milagre have incomes of less than one salary (Table 5.4).

TABLE 5.4 - **Income indicator values for sub-basin 1.**

INCOME INDICATOR - IR	
NEIGHBORHOODS	**INDICATOR**
Center	1

Mother of God	0,25
Apicum	1
Desterro	0
Goiabal	0,25
Lira	0,25
Vila Passos	0
Monte Castelo	1
Sand	0,25
Good Miracle	0
Parque Amazonas	1
Fâtima neighborhood	0,25
Apeadouro	1
ToTAL	6,25
INDICATOR AVERAGE	0,48

Camargo Mora (2003), investigating the determinants of environmental values between countries, observed that economic wealth has a direct influence on environmental awareness. The article shows that the degree of urbanization, the level of subjective well-being and the level of income equality have a direct effect on awareness, while education, demographic pressure and happiness are significantly correlated with environmental behaviour.

5.1.2.2 Education Indicator (IE)

Biological and physical aspects constitute the natural basis of the human environment, while socio-cultural and economic dimensions and ethical values define the guidelines and tools with which man can better understand and use nature's resources (sustainability) in order to satisfy his needs (DIAS, 2004).

According to the education indicator (IE) with a value of 0.47 (Table 5.5), sub-basin 1, located in the commercial center of the city of São Luis, has a reasonable number of schools, showing easy access for this population.

TABLE 5.5 - **Education indicator values for sub-basin 1.**

EDUCATION INDICATOR - IE

NEIGHBORHOODS	INDICATOR
center	0,7
Mother of God	0,3
Apicum	1
Desterro	0,3
Goiabal	0,3
Lira	0,3
Vila Passos	0,7
Castle Hill	0,7
Sand	0,3
Good Miracle	0,3
Parque Amazonas	0,3
Fàtima neighborhood	0,3
Apeadouro	0,7
ToTAL	6,2
INDICATOR AVERAGE	0,47

The Desterro and Bom Milagre neighborhoods have the lowest income levels. This reality directly implies a low level of schooling, since the lack of monetary resources "forces" most people to drop out of school to join the informal labor market in order to help with the family income. Combating inequality in income distribution is fundamental to ensuring poverty reduction, one of the main challenges of sustainable development.

Portilho (2009) mentions that the level of environmental awareness of a people can be assessed by the development of an integrated understanding of the environment in its multiple and complex relationships, involving physical, biological, social, cultural, economic, scientific and ethical aspects.

5.1.2.3 Sanitation Indicator (SI)

The value for the sanitation indicator, 0.95 (Table 5.6), shows that this sub-basin has water supply and sewage network services. On the other hand, only 15% of the sewage collected is treated (with staff), and the rest is deposited *in natura in* the Bacanga river basin.

TABLE 5.6 - **Sanitation indicator values for sub-basin 1.**

SANITATION INDICATOR - IS		
NEIGHBORHOODS	WATER	SEWAGE
Center	1	1
Mother of God	1	1
Apicum	1	1
Desterro	1	1
Goiabal	1	0,7
Lira	1	0,9
Vila Passos	1	1
Monte Castelo	1	1
Sand	1	0,85
Good Miracle	1	1
Parque Amazonas	1	1
Fàtima neighborhood	1	1
Apeadouro	1	1
INDICATOR AVERAGE	1	0,95

The neighborhoods with the lowest levels of education and income were the same as those with the lowest levels of sanitation, thus demonstrating a very close relationship. The higher the level of education, the greater the population's demand for a better quality of life.

The Socioeconomic Dimension (SD) showed an index of 0.63. This sub-basin enjoys essential services such as schools, water supply and sewage systems.

The factors that impact on the sustainability of water resources in a river basin are related to social, economic and ecological aspects.

Thus, the ISAsub1 found was 0.43. This median value is due to the fact that the ED has a very low value, due to the irregular growth of the urban population, which has caused strong impacts affecting the soil, vegetation, water quality and, consequently, man himself.

5.2 Sub-basin 2

Like sub-basin 1, sub-basin 2 also has old occupation, with disordered growth and populous neighborhoods such as coroadinho. Its main river is the Bicas, one of whose sources is located in the Salinas do Sacavém neighborhood, *in* the middle of houses from which sewage flows into the river.

5.2.1 Ecological Dimension (ED)

5.2.1.1 Physico-chemical variable (PV)

Like sub-basin 1, sub-basin 2 only showed a pH value in accordance with CONAMA 357/05, with a value of 7.54, while the other variables showed values outside the proposed standard (Table 5.7).

According to Macêdo (1978), in estuarine regions, pH is related to the amount of decomposing organic matter, which is a limiting factor for both the mixing of carbon dioxide and dissolved oxygen.

Nutrient levels are often related to the degree of pollution in an aquatic ecosystem, and when they are high, they are generally interpreted as indicators of a polluted environment, demonstrating an advanced state of eutrophication (CARMOUZE, 1994). In sub-basin 2, the nitrate values of 0.87 mg/L proved to be above the CONAMA 357/05 standard value, as was the case for ammonia with 1.81 mg/L and phosphate 4.67μM. These values show a more pronounced state of pollution than that seen in sub-basin 1 (Table 5.7).

TABLE 5.7 - **Values of the physico-chemical variables for sub-basin 2, average value of the samples, CONAMA 357/05 reference value, indicator value per variable and average of the indicators.**

PHYSICAL-CHEMICAL VARIABLES - VF			
VARIABLES	**AVERAGE VALUE X± SD**	**CONAMA 357/05**	**INDICATOR VALUE**
pH	7,54 (±0,78)	6,5 - 8,5	1
OD	2.27 (±1.00) mg/L	≥ 5 mg/L	0
AMON	1.81 (±1.26) mg/L	≤ 0.4 mg/L	0
NITRATE	0.87 (±1.06) mg/L	≤ 0.4 mg/L	0
PHOSPHATE	4.67 (±4.63) μM	1.2 μM	0
INDICATOR AVERAGE			**0,2**

All forms of phosphorus are important, but orthophosphates are the most important because they are the main form of phosphate assimilated by aquatic plants (CHESTER, 1990).

In NE estuaries, anthropogenic emissions of N and P are 20 to 50 times greater than natural inputs via the soil and atmosphere (MARINS *et al.*, 2003).

5.2.1.2 Biological Variable (BV)

For the indicator of the biological variable, a value of 0.75 was obtained, showing greater diversity than that found in sub-basin 1 (Table 5.8). In estuarine waters, the predominance of one group or another depends mainly on the dominant characteristics of the environment, and the structure of the phytoplankton community is an important parameter for assessing water quality (ESKINAZI-LEÇA *et al.*, 2002).

Labohidro (1998), in a study carried out in the Bacanga River estuary, identified 34 taxa, distributed in two divisions: Bacillariophyta (91%) and Cyanophyta (9%). The Bacillariophyta division has a preference for areas rich in nutrients, supporting greater variation, *while the* presence of cyanobacteria is a sign of dulcicolous influence in the estuary.

TABLE 5.8 - **Values of the biological variable per collection (density - Shanonn Diversity Index), value per indicator per collection and average of the indicators for sub-basin 2.**

BIOLOGICAL VARIABLE - PHYTOPLANKTON		
COLLECTIONS	**SHANNON DIVERSITY INDICE (nits.cel)$^{-1}$**	**INDICATOR VALUE**
1	1,26	0,3
2	2,45	1
3	2,43	1
4	1,71	0,7
INDICATOR AVERAGE		**0,75**

According to Paula Filho (2004), algal blooms are the result of an increase in nutrients in water sources, producing certain effects on water quality. He discusses the relationship between the phytoplankton community and environmental variations, which are essential for understanding these environments. Algae are influenced by natural and artificial environmental conditions. They vary in quantity and frequency according to various factors, such as climate, water temperature, seasons and anthropogenic influences.

According to Labohidro (1998), the Bicas river, the main part of sub-basin 2, is exposed to environmental degradation and is considered the most polluted.

According to Margalef (1983), specific diversity values ranging from 1 to 2.5 bits/cell characterize the coastal region and can be especially low in estuaries, early successional communities and polluted environments. Values between 3.5 and 4.5 bits/cell are found in oceanic regions or in coastal communities in more advanced stages of succession.

5.2.1.3 Environmental Preservation Index (IP)

Although sub-basin 2 has a significant vegetated area, the anthropized areas give it a higher value, showing the various activities carried out by man.

This basin is surrounded by populous neighborhoods such as Coheb, Coroado and Pòlo Coroadinho, whose paved areas have contributed to less surface runoff from the sub-basin. The IP (0.40) (Table 5.9) of this sub-basin showed a marked level of degradation.

TABLE 5.9 - **Environmental Preservation Index (IP) values for sub-basin 2.**

ENVIRONMENTAL PRESERVATION INDEX - IP	
LAND USE CLASSES	**VALUES**
Drainage Density (Dd)	2.55 Km/ Km^2
Vegetated Area (VA)	7.81 Km^2
Soaked Areas (AE)	1.49 Km^2
Anthropized Areas (AA)	8.99 Km^2
Environmental Preservation Index (IP)	**0.40** Km^2

The occupation of these areas is totally illegal and represents a social and economic risk for these families, since they are located in an area that presents a risk of flooding, as this area is situated in one of the lowest parts of the city.

Alcântara and Amorim (2005) carried out a study in the Anil river basin and found that from 1980 to 1999 the Anil river drainage basin was intensely modified by the growing urbanization in the basin, an event also seen in the Bacanga river basin.

In this sub-basin, it can also be seen that during the rainy season, flooding occurs on the avenue that gives access to these neighborhoods (Avenida Médice) and on many streets in Coroado and Coroadinho, demonstrating poor urban drainage.

The flooded area (Figure 5.2) corresponds to the Bicas River, the main tributary of this sub-basin, which receives domestic effluent and has very high levels of phosphate and ammonia, as well as low levels of DO.

Thus, the ecological dimension (ED) showed a value of 0.45, different from the value for sub-basin 1 (ED = 0.24). This difference refers to the Environmental Preservation Index (IP), which is higher in sub-basin 2 than in sub-basin 1. This is due to the fact that sub-basin 1 has a large anthropized area and little vegetation, while sub-basin 2, despite having high levels of pollution in the Bicas River, influencing the physico-chemical variables, still has an extensive area of vegetation.

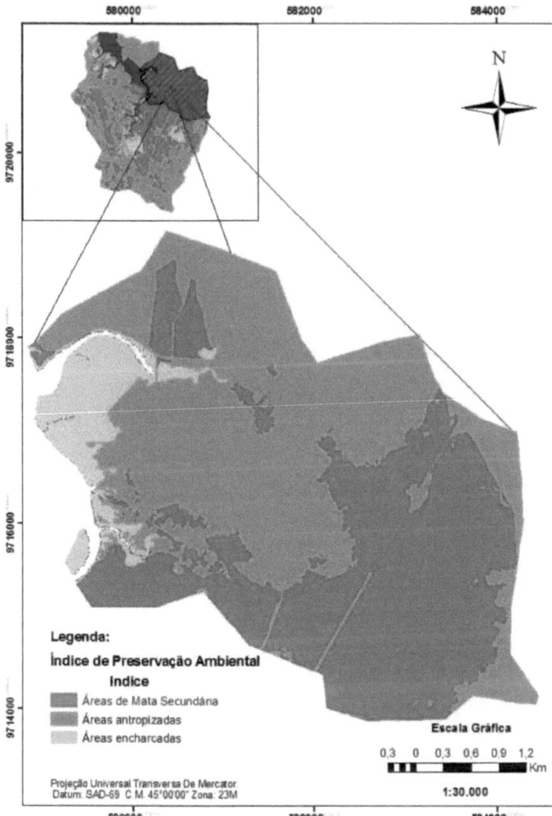

FIGURE 5.2 - **Graphical representation of the Environmental Preservation index (IP) for sub-basin 2.**

5.2.2 Socioeconomic Dimension (SD)

The population of this sub-basin is represented by an income below one salary, an amount considered insufficient for a family to survive, characterizing a low-income population (IPEA, 1998).

5.2.2.1 Income Indicator (IR)

In this sub-basin, the neighborhoods with incomes above five minimum wages were Filipinho, Sítio Leal and Pindorama, while the others were between one and one minimum wage (Table 5.10).

TABLE 5.10 - **Income indicator values for sub-basin 2.**

INCOME INDICATOR - IR	
NEIGHBORHOODS	**VALUE**
Jordoa	0
Crowned	0,25
Filipinho	1
Sitio Leal	1
Salina Sacavém	0
Sacavém	0,25
Coheb	0
Outeiro da Cruz	0,25
Sitio Pirapora	0,25
Vila dos Frades	0
Coroadinho	0
Timbira Park	0,25
Dom Sebastiao Complex	0
Vila dos Nobres	0
Pindorama	1
Spring	0
Vila Conceiçao	0
TOTAL	4,25
INDICATOR AVERAGE	0,25

Disordered urban growth, often due to rural exodus or a high birth rate, generates the appearance of invasions, which are characterized by a low-income population living in places without infrastructure and welfare services. This population generally does not have a qualified workforce and survives on informal services.

There is an association between the socio-economic position of the population and fertility levels, in the sense that the less educated groups still have higher fertility rates, while the intermediate and more educated groups have lower rates, as they have the means and information to plan their families, and are more aware of their socio-economic and environmental responsibilities (IBGE, 2009).

5.2.2.2 - Education Indicator (IE)

The Sitio Leal and Pindorama neighborhoods (Table 5.11) had the highest values for education and income. These areas originate from subdivisions intended for the construction of housing estates for the upper middle class.

TABLE 5.11- **Education indicator values for sub-basin 2.**

EDUCATION INDICATOR - IE	
NEIGHBORHOODS	**VALUE**
Jordoa	0,3
Crowned	0,3
Filipinho	0,3
Sitio Leal	1
Salina Sacavém	0,3
Sacavém	0,3
Coheb	0,3
Outeiro da Cruz	0,3
Sitio Pirapora	0,3
Vila dos Frades	0,3
Coroadinho	0,3
Timbira Park	0,3
Dom Sebastião Complex	0,3
Vila dos Nobres	0,3
Pindorama	0,7
Spring	0,3
Vila Conceição	0,3
TOTAL	6,2

INDICATOR AVERAGE	0,36

Education is fundamental to environmental awareness. The role of education in the educational process is emphasized in Chapter 36 of AGENDA 21 - Promoting Education, Awareness and Training: education is of fundamental importance in promoting sustainable development and increasing people's capacity to address environmental and development issues. Both formal and informal education are indispensable for changing people's attitudes, so that they have the capacity to assess and address the problems of sustainable development.

5.2.2.3 -Sanitation Indicator (SI)

More than 50% of the neighborhoods (Table 5.12) in this sub-basin had a value of zero for the sewage variable, which is due to the fact that these neighborhoods are invasions of the Coroadinho pole.

TABLE 5.12 - **Sanitation indicator values for sub-basin 2.**

SANITATION INDICATOR	AMOUNT - IS	
NEIGHBORHOODS	**WATER**	**SEWAGE**
Jordoa	1	0,9
Crowned	1	0
Filipinho	1	1
Sitio Leal	1	0,95
sacavém saltworks	1	0
sacavém	1	0,2
Coheb	1	1
Hill of the Cross	1	0,6
Vila Conceição	0,75	0
Vila dos Frades	0,75	0
Coroadinho	1	0,7
Timbira Park	1	1
Conjunto Dom Sebastiao	0,75	0
Vila dos Nobres	1	0,8

Pindorama	1	0
Spring	0,25	0
Sitio Pirapora	0,5	0
AVERAGE INDICATORS	0,88	0,42

In addition to the lack of sewage and irregular water supply, the image of this sub-basin is usually associated with the poverty of its population, most of whom are excluded from the system due to society's prejudice against their socio-economic and cultural conditions, thus contributing to the high unemployment rate.

The precarious conditions observed in this sub-basin are characterized by a lack of infrastructure and essential services, such as an incipient sewage system. Drainage is almost always in the open and, although most of the houses have toilets, most of the human waste goes into septic tanks or "black cesspools" (FROES, 2004).

The Socioeconomic Dimension (SD) of sub-basin 2 is represented by a value of 0.42. The ISAsub 2 value is represented by the same value as sub-basin 1 (0.43). Peripheral regions are the result of the accelerated urbanization processes that mark modern societies and are the bearers of numerous structural, social and human problems, with consequences that go beyond their geographical limits and affect the city as a whole. The two sub-basins have undergone the same process of disorganized population expansion, especially sub-basin 1, which was occupied the longest.

The region known as Pólo Coroadinho emerged around 1976, forming on the banks of the Bicas River. Most of the families came from the interior of the state in search of better living conditions.

After the occupation, the population expanded to other areas, giving rise to 16 (sixteen) villages and agglomerations that make up the pole, which are: Coroadinho, Vila dos Frades, Bom Jesus, Primavera, Sitio do Pica Pau Amarelo, Vila Conceiçâo, Vila dos Nobres, Vila Natal, Vila Atêncio Queiroga, Vila Sâo Sebastiâo, Terra Preta, Alto da Bela Vista, Alto do Sâo Francisco, Alto do Parque Timbiras, Vila Maruim and Vila Nice Lobâo. (RIBEIRO JÚNIOR, 1999)

5.3 Sub-basin 3

Sub-basin 3 is part of the Bacanga State Park, which, as a Conservation Unit (UC), represents the Environmental Heritage of the island of São Luis. This sub-basin has two neighborhoods: Vila Itamar and the Industrial District, as well as the Recanto Verde invasion.

5.3.1 Ecological Dimension (ED)

5.3.1.1 Physico-chemical variable (PV)

In this sub-basin, it was found that only the physico-chemical variables DO and phosphate were not within the standard required by CONAM Resolution A 357/05. The low oxygen values, 3.51 mg/L (Table 5.13) can be attributed to the greater presence of organic load from the dumping of domestic waste from the urban occupations of Vila Itamar and Recanto Verde.

TABLE 5.13 - **Values of the physico-chemical variables for sub-basin 3, average value of the samples, CONAMA 357/05 reference value, indicator value for the variable and average of the indicators.**

PHYSICAL-CHEMICAL VARIABLES - VF			
VARIABLES	**AVERAGE VALUE X ± SD**	**CONAMA 357/05**	**VALUE OF INDICATOR**
pH	7,32 (±0,99)	6,5 - 8,5	1
OD	3.51 (±1.72) mg/L	≥ 5 mg/L	0
AMON	0.14 (±0.14) mg/L	≤ 0.4 mg/L	1
NITRATE	0.37 (±0.45) mg/L	≤ 0.4 mg/L	1
FOsFACT	1.46 (±1.13) μM	1.2 μM	0
AVERAGE OF INDICATORS			**0,6**

The values for ammonia (0.14), nitrate (0.37) and pH (7.32) were within the limits of CONAMA Resolution 357/05, unlike the values observed for the other sub-basins studied, demonstrating a lower degree of pollution (Table 5.13).

Cavalcante (2005) carried out a study on the use and occupation of the Batatâ sub-basin and the effects on the degradation of its water resources, finding low levels of oxygen at various sampling points, with values ranging from 0.7 to 2.9 mg/L, indicating that these waters are compromised.

Phosphorus is an essential nutrient for living organisms and its presence at optimal levels favors the development of algae. Natural sources of phosphorus can be found in domestic sewage discharges, industrial effluents and leaching from agricultural areas, among others (GASTALDINI *et al.*, 2003).

5.3.1.2 - Biological Variable (BV)

Water pollution puts an end to various forms of life that need it, both plant and animal, from phytoplankton and zooplankton to the largest creatures that inhabit waterways, fish, as well as affecting human beings who are obligatory consumers of water from these environments. Changes in environmental conditions lead to a restructuring of the fish assemblages, reflecting the conditions prevailing in the watershed in which they are located (MARTINS, 2008).

The value for the biological variable was the lowest found compared to sub-basins 1 and 2 (Table 5.14). In estuaries, diversity can be low due to the high degree of eutrophication. When nutrient concentrations are high, there is a greater presence of r-strategist or opportunist species of small size and with a high reproduction rate, increasing algal biomass and decreasing diversity (MARGALEF, 1983).

TABLE 5.14 - **Values of the biological variable per collection (density - Shanonn Diversity Index), value of the indicator per collection and average of the indicators for sub-basin 3.**

BIOLOGICAL VARIABLE - PHYTOPLANKTON		
COLLECTIONS	**DIVERSITY INDEX DE SHANNON (nits.cel)$^{-1}$**	**VALUE OF INDICATOR**
1	1,32	0,3
2	1,08	0,3
3	0,69	0
4	2,38	1
INDICATOR AVERAGE		**0,4**

5.3.1.3 - Environmental Preservation Index (IP)

Pereira (2003), in an "in loco" survey of this sub-basin, found a high degree of anthropism and environmental degradation in the drainage network, in the part of it that includes the residential areas of Vila Itamar and Recanto Verde. There has been a significant de-characterization of the natural setting, through points of intermittence in the drainage network, due to the triggering of erosive processes, silting and deposits of solid domestic waste distributed along the channels.

The vegetated areas (20.99 km^2) are higher than in the other sub-basins studied (Table 5.15). This is due to the fact that this sub-basin is home to the Bacanga State Park, a Conservation Unit, and the Batatâ reservoir, which is responsible for supplying water to several neighborhoods in São Luis.

TABLE 5.15 - **Environmental Preservation Index (IP) values for sub-basin 3.**

ENVIRONMENTAL PRESERVATION INDEX - IP	
LAND USE CLASSES	**VALUES**
Drainage Density (Dd)	2.90 Km/ Km^2
Vegetated Area (VA)	20.99 Km^2
Soaked Areas (AE)	2.24 Km^2
Anthropized Areas (AA)	15.25 Km^2
Environmental Preservation Index (IP)	**0.52** Km^2

The Bacanga State Park was the first government initiative to preserve ecological balance and determine the use of water resources.

In 1984, Decree No. 9550, of April 10, established new boundaries for the Conservation Unit (UC), increasing it from 3,075 ha to 2,634 ha. This redefinition was due to some irregular occupations, exploitation of natural resources, as well as the installation of the Sâo Luis - Teresina railroad and Eletronorte's transmission lines, leading to serious environmental impacts on the existing natural ecosystems and the species found within it (CAVALCANTE, 2005).

Unfortunately, the municipalities have little capacity for oversight, and end up developing in the source areas with irregular or clandestine allotments. The graphical representation (Figure 5.3) of the IP for this sub-basin showed a large anthropized area near the Batatâ lakes. The consequence of expansion without an environmental vision is the deterioration of the water sources and the reduction of safe water coverage for the population. This process requires a variety of preventive urban and environmental planning actions, aimed at minimizing impacts and seeking sustainable development.

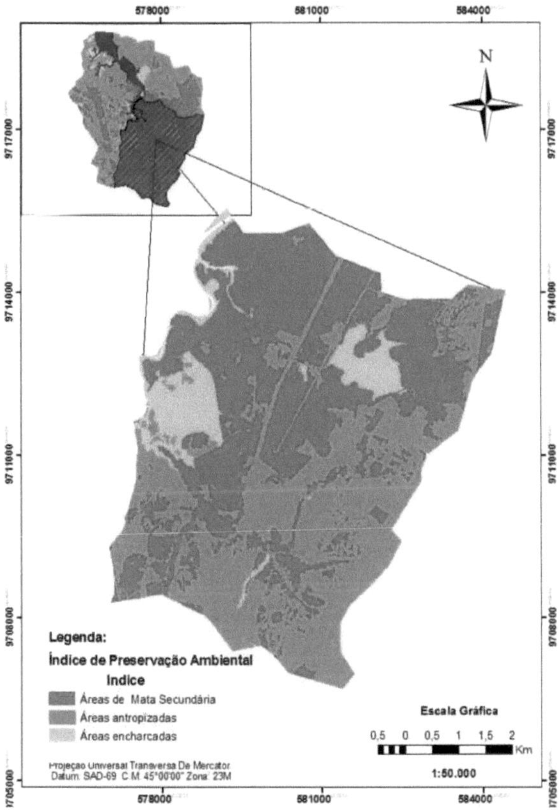

FIGURE 5.3 - **Graphical representation of the Environmental Preservation index (IP) for sub-basin 3.**

5.3.2 Socioeconomic Dimension (SD)

The process of creating the Vila Itamar and Distrito Industrial neighborhoods has similar origins. The national decision to set up major projects in Sâo Luis, such as the Alumar aluminium factory, the Itaqui export port and the Carajás railroad owned by Companhia Vale do Rio Doce, attracted countless service companies and helped to consolidate the idea of an imminent housing explosion (BURNETT, 2006)

In 1964, families of employees of the Department of Roads and Highways (D.E.R.) began to occupy the area that served as a garage for that institution, in the vicinity of the BR 135 highway, close to the runway of the Marechal Cunha Machado airport (PEREIRA, 2003).

These neighborhoods and later the Vila Recanto Verde (invasion), contributed to the de-characterization and degradation of the UC.

Social facilities such as schools are insufficient for the demand in the area. The main economic activities are informal services such as market traders, bricklayers, carpenters, tradesmen, electricians, drivers, domestic workers, fishermen and bag pickers (PEREIRA, 2003).

Education is fundamental to providing environmental and ethical awareness, values and attitudes, techniques and behaviors that are in line with sustainable development and that encourage effective public participation in decision-making.

A population without educational training will never demand their rights to basic services (water supply, sewage disposal, garbage collection) and neither will they have the awareness to examine their attitudes towards natural resources.

5.3.2.1 - Income Indicator (IR)

The income indicator (0.12) was very low compared to the other sub-basins (Table 5.16). The Vila Itamar neighborhood had a population living on salaries below the minimum wage.

TABLE 5.16 - **Income indicator values for sub-basin 3.**

INCOME INDICATOR	**- IR**
NEIGHBORHOODS	**VALUE**
Industrial District	0,25
Vila itamar	0
INDICATOR AVERAGE	0,12

The low qualification of the workforce is usually a limiting factor in the possibilities of obtaining higher incomes.

Education is not just a question of citizenship, but also of public health, which is directly related to productivity. A poor family without sanitation is more likely to suffer from diseases, especially malnourished children.

Faced with the intense need to live in the city, a strong attack on the environment is unleashed, with negative effects on the watercourses, which begin to suffer the consequences of a socio-educational problem, resulting from the lack of housing and insufficient environmental education practice to sensitize the migrant population to conserve these environments.

5.3.2.2 - Education Indicator (IE)

The average of the education indicator, 0.30 (Table 5.17) shows that the population only has primary education, but that this is working poorly in terms of the teaching-learning process, compromising the basis for the formation of citizens. At this level of education, there is a predominance of supplementary education to make up for lost years and make time available for work.

TABLE 5.17 - **Education indicator values for sub-basin 3.**

EDUCATION INDICATOR - IE	
NEIGHBORHOODS	**VALUE**
Industrial District	0,3
Vila Itamar	0,3
INDICATOR AVERAGE	0,3

The historical and early participation of poor young workers reveals not only how Brazilian society has been configured throughout its history, but also the aggravation that low-income families experience as a result of the increasingly recurrent entry of their children into precarious, marginal work and even drug trafficking activities, in order to make up the family income. Even in activities where a minimum of schooling and knowledge prevails, there is a relationship of incompatibility between work and school. It's a reality that drives their arms and bodies to the school benches at night, immersed in tiredness, sleep, lack of concentration and, above all, the absence of time to dedicate to the little knowledge produced in the classroom.

5.3.2.3 Sanitation Indicator (SI)

The availability of water for the neighborhoods in this sub-basin reaches only 42% of the residents of Vila Itamar and 45% for the Industrial District (Table 5.18). It should be noted that in Vila Itamar, a large part of the existing area is currently occupied by squatters.

TABLE 5.18 - **Sanitation indicator values for sub-basin 3.**

SANITATION INDICATOR - IS		
NEIGHBORHOODS	**WATER**	**SEWAGE**
Industrial District	0,45	0,2
Vila Itamar	0,42	0,13
INDICATOR AVERAGE	0,43	0,16

The DS = 0.23 result for this sub-basin demonstrates the lack of commitment of public bodies to the community; urban agglomerations with no infrastructure, lack of schools, irregularities in the water supply and the absence of a sewage system, reflecting on the local environmental quality.

ISAsub 3 was 0.36, lower than in the two sub-basins studied above, meaning that this sub-basin is at an accelerated stage of degradation.

5.4 Sub-basin 4

This sub-basin is located in the Itaqui-Bacanga area, on the left bank of the Bacanga River, and is made up of the Vila Embratel and Gapara neighborhoods. The process of occupation of this area is associated with the construction of the Bacanga river dam in the 60s.

5.4.1 Ecological Dimension (ED)

5.4.1.1 Physico-chemical variable (PV)

For the physico-chemical variables, the OD (2.93) showed a low value, which may be associated with the effluents that contribute a large quantity of organic compounds to the water bodies, leading to the process of eutrophication (Table 5.19).

Eskinazi-Leça *et al.* (1984), investigating the Botafogo River (PE), which has been suffering from anthropogenic action, found OD values between 1.63 and 4.73 mg/L.

The other variables (ammonia, nitrate and phosphate) also obtained values higher than those set out in CONAMA Resolution 357/05. The pH value was within the proposed limit.

The average value of the physico-chemical variables was 0.2 (Table 5.19), the same result as that found for sub-basins 1 and 2. It should be noted that sub-basin 4 is an area of recent urban expansion, showing that this sub-basin is undergoing a rapid process of degradation.

TABLE 5.19 - **Values of the physico-chemical variables for sub-basin 4, average value of the samples, CONAMA 357/05 reference value, indicator value for the variables and average of the indicators.**

PHYSICAL-CHEMICAL VARIABLES - VF			
VARIABLES	**AVERAGE VALUE X ± SD**	**CONAMA 357/05**	**VALUE OF INDICATOR**
pH	7,06 (± 1,39)	6,5 - 8,5	1
OD	2.93 (±0.63) mg/L	≥ 5 mg/L	0
AMONIA	1.06 (±0.91) mg/L	≤ 0.4 mg/L	0
NITRATE	1.35 (±0.98) mg/L	≤ 0.4 mg/L	0
PHOSPHATE	1.89 (±3.02) μM	1.2 μM	0

AVERAGE INDICATORS	0,2

According to MOTA (1995), high concentrations of nitrogen compounds (ammonia, nitrites, nitrates, among others) can be used as indicators of water quality.

5.4.1.2 Biological Variable (BV)

The diversity of phytoplankton organisms found in this sub-basin was 0.67 (Table 5.20). Various proposals have been put forward to explain the diversity of planktonic species as being the result of the interaction of factors such as competition, predation, size, habitat, time and environmental stability (KREBS, 1989).

TABLE 5.20 - **Values of the biological variable per collection (density - Shanonn Diversity Index), value of the indicator per collection and average of the indicators for sub-basin 4.**

BIOLOGICAL VARIABLE - PHYTOPLANKTON		
COLLECTIONS	SHANNON DIVERSITY INDICE (nits.cel-1)	INDICATOR value
1	2,07	1
2	1,99	0,7
3	1,42	0,7
4	1,20	0,3
INDICATOR AVERAGE		0,67

Planktonic organisms establish complex intra- and inter-specific relationships, competing for space and organic and inorganic resources in the water column. Changes in the planktonic community lead to profound structural changes in all trophic levels of aquatic ecosystems because, due to its dynamic nature, with high rates of reproduction and loss, this community responds quickly to physical and chemical changes in the environment.

Variations in the meteorological regime, regional geomorphological characteristics and anthropogenic impacts on coastal and estuarine areas together establish the particular hydrographic regime of each region and, consequently, the taxonomic characteristics and spatio-temporal dynamics of its communities (SANTOS, 2009).

5.4.1.3 Environmental Preservation Index (IP)

Drainage density is an important factor in indicating the degree of development of a basin's drainage system. As such, this index provides an indication of the basin's drainage efficiency

(ANTONELI; THOMAZ, 2007).

The basin under study has an average drainage density of 2.50 km/km^2 (Table 5.21); this value is a reflection of a relief with greater topographic movement and a shorter concentration time (VILLELA; MATTOS, 1975). The vegetated areas showed a relatively low value when compared to the anthropized area (Figure 5.4). This is due to the rapid increase in population in this sub-basin since the 60s.

TABLE 5.21 - **Environmental Preservation Index (IP) values for sub-basin 4.**

ENVIRONMENTAL PRESERVATION INDEX - IP	
LAND USE CLASSES	**VALUES**
Drainage Density (Dd)	2.50 Km/ Km^2
Vegetated Area (VA)	11.97 Km^2
Soaked Areas (AE)	1.44 Km^2
Anthropized Areas (AA)	9.48 Km^2
Environmental Preservation Index (IP)	**0.56** Km^2

Due to unplanned urban growth, some basins have high levels of waterproofing, which has led to an increase in runoff volumes, flooding, erosion and siltation of the macro-drainage. This agrees with Tucci's (2000) assertions about the relationship between an increase in impermeable area and an increase in the flood coefficient and the volume of surface runoff.

Waterproofing and soil compaction, as well as siltation caused by the removal of vegetation, cause serious damage, since the influence of freshwater flow, tides and winds results in specific patterns of water circulation in the estuary. Christofoletti (1979) assumes that the reduction in freshwater inflow is the most serious physical effect generated by changes in the drainage basin, since it intensifies the worsening of pollution by reducing estuarine circulation.

The ED value for this sub-basin is 0.47, close to the value found in sub-basin 2. As they stand, the problems are worsening and cities are becoming the scene of uncontrolled and disorganized action, both in terms of the poor distribution and lack of control of frequent urbanization in space, and environmentally, since urbanization without efficient urban planning reflects a joint socio-economic crisis.

Among the factors that lead to environmental transformation are deforestation and the occupation of the banks of urban watercourses for housing purposes, which are factors of environmental degradation, accelerating the process of soil erosion and consequent silting up of the streams, thereby reducing the quality of life of the population that lives there.

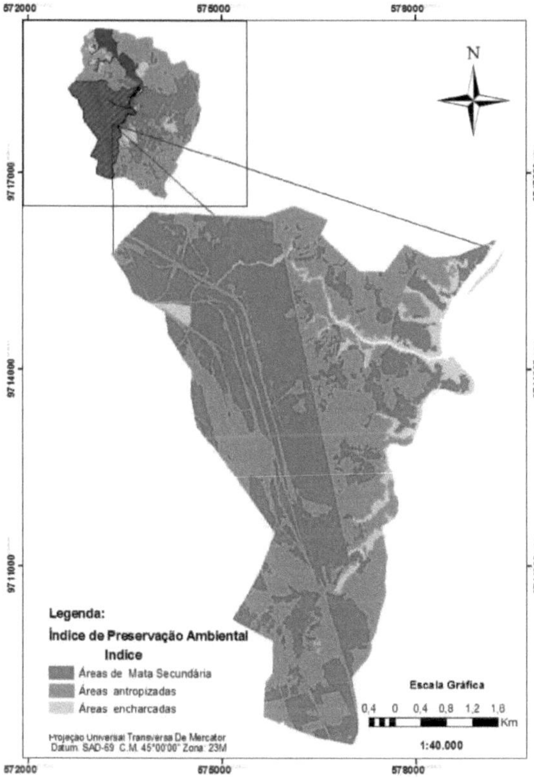

FIGURE 5.4 - **Graphical representation of the Environmental Preservation index (IP) for sub-basin 4.**

5.4.2 Socio-economic Dimension (SD)

Due to a series of factors, mainly social and economic, the number of irregular occupations continues to grow, also increasing urban problems, especially those of a metropolitan nature.

Certain concepts should be reviewed as to what constitutes irregular occupation, based on socio economic, cultural and environmental characteristics, since there is a great deal of socio-environmental instability made up of a deprived population, usually located in areas of environmental sensitivity.

5.4.2.1 Income Indicator (IR)

A large part of the population living in these areas comes from the interior of the state, attracted by the expectation of employment and a better life, created by the establishment of industries in the

Itaqui-Bacanga area, which ended up causing greater demand than supply of work, resulting in unemployment.

The IR value (0.12) in this sub-basin was also found for sub-basin 3. The Vila Embratel and Gapara neighborhoods have a low-income population that lives off informal services, often having to abandon their studies to contribute to the family budget (Table 5.22). Another option for these families comes from the government through the Bolsa Familia welfare program.

TABLE 5.22 - **Income indicator values for sub-basin 4.**

INCOME INDICATOR - IR	
NEIGHBORHOODS	**VALUE**
Vila Embratel	0,25
Gapara	0
INDICATOR AVERAGE	**0,12**

Vila Embratel is home to a UFMA Extension Center that works on various projects for the community, including vocational courses, giving the population an opportunity to improve their quality of life. However, what we see is a lack of interest on the part of those who really need an alternative source of income.

5.4.2.2 Education Indicator (IE)

The urban landscape is a complex system in which different elements and components represent flows of energy and matter in interactive processes. Understanding the urban landscape as a system can lead to analysis, classification and a synthetic diagnosis. The elements that will dominate these landscapes are anthropic, although biotic and abiotic elements also play a role. Anthropic energy undoubtedly prevails over natural energy (SILVA, 2009).

Social inequality is usually analyzed in terms of the differences between the upper and lower classes and thought of above all as inequality of income and education. It should be noted that not all people on low incomes are the same and that it is important to recognize the existence of inequalities within a class. In other words, although income and education are central dimensions in social differentiation, the spatial dimension of inequality should not be neglected, as it manifests itself in different forms of socio-spatial segregation (PRETECEILLE; RIBEIRO, 1999).

This fact can be seen in these neighborhoods, since within their boundaries there is a physical stratification, i.e. there are places with paved streets, close to schools, squares, the market and the church. On the other hand, there are marginalized areas such as América do Norte, Piancor and

Residencial Resende, which are part of Vila Embratel but do not have the same infrastructure.

For the EI, the result was 0.3 (Table 5.23), which shows that this population has only primary schooling. It can be inferred that education has the impact of perpetuating the cycle of poverty, since parents with low levels of education find it difficult to guarantee a higher level of education for their children, thus generating a vicious cycle of perpetuating poverty between generations.

TABLE 5.23 - **Education indicator values for sub-basin 4.**

EDUCATION INDICATOR	**- IE**
NEIGHBORHOODS	**VALUE**
Vila Embratel	0,3
Gapara	0,3
INDICATOR AVERAGE	**0,3**

5.4.2.3 Sanitation Indicator (SI)

The lack of infrastructure and services has made a large part of the population in this sub-basin increasingly vulnerable to disease, revealing a profound social inequality caused by the process of segregation of the majority of the population to the most environmentally fragile (areas known as baixios, with potential for flooding or on the slopes), unhealthy and deteriorated portions, creating a geography of exclusion with profound differences in the quality of housing, access to goods and services (Table 5.24). The absence of sewage services, as seen in the other sub-basins, is present in the neighborhoods of this sub-basin, especially in Gapara. The main consequence is the contamination of surface and underground water sources. This demonstrates the lack of investment in basic services for the population and disregard for the environment.

TABLE 5.24 - **Sanitation indicator values for sub-basin 4.**

SANITATION INDICATOR		**- IS**
NEIGHBORHOODS	**WATER**	**SEWAGE**
Vila Embratel	0,80	0,10
Gapara	0,50	0
INDICATOR AVERAGE	0,65	0,05

The SD for this sub-basin was 0.25, representing a very low value for issues relating to the population's socio-economic development, which is reflected in citizens' environmental awareness.

The ISAsub for sub-basin 4 showed the same value as sub-basin 3, demonstrating that the environmental socio-economic situation is the same for the two sub-basins.

From the perspective of sustainable development, population growth must be accompanied by the conditions for satisfying human needs, such as combating poverty and reducing unemployment, thus minimizing the impact on the environment.

5.5 Sub-basin 5

As in sub-basin 4, the occupation process in this sub-basin was directly related to the construction of the dam. As well as providing access to the left bank of the Bacanga river, allowing urban occupation, the dam also became an obstacle to the ebb and flow of tidal waters, altering the natural hydrodynamics of the estuarine system and the volume of water circulating and residing in the system. This sub-basin is home to one of the most populous neighborhoods in the Itaqui-Bacanga region, Anjo da Guarda.

5.5.1 Ecological Dimension (ED)

5.5.1.1 Physico-chemical variable (PV)

The indicator values for the physico-chemical variables for this sub-basin are the same as those found in sub-basins 1, 2 and 4. This pattern reflects anthropogenic actions on water quality.

It was found that the nitrate level (1.37) in this sub-basin was the highest recorded when compared to the other sub-basins (Table 5.25). Nitrogen exported through point and diffuse sources can result in an increase in primary productivity, which induces an acceleration in the eutrophication process of water bodies, with the potential seasonal development of toxic algae, which can have major impacts on overall water quality (ESCHRIQUE, 2007).

TABLE 5.25 - **Values of the physico-chemical variables for sub-basin 5, average value of the samples, CONAMA 357/05 reference value, indicator value for the variable and average of the indicators.**

PHYSICAL AND CHEMICAL VARIABLES			
VARIABLES	AVERAGE VALUE X ± SD	CONAMA 357/05	INDICATOR VALUE
pH	7,32 (±1,08)	6,5 - 8,5	1
OD	3.51 (±1.14) mg/L	≥ 5 mg/L	0
AMONIA	0.72 (±0.46) mg/L	≤ 0.4 mg/L	0

NITRATE	1.37 (±**1**.19) mg/L	≤ 0.4 mg/L	0
PHOSPHATE	2.37 (±**3**.29) μM	1.2 μM	0
AVERAGE INDICATORS			**0,2**

Natural eutrophication occurs quite slowly, through the input of nutrients brought by rainfall, surface water and resuspension processes. The artificial acceleration of this process is mainly due to the volume of nitrogenous and phosphate inputs from domestic and industrial effluents and agricultural activities (BENDO *et al.* 2000).

Nitrates are very soluble in water and are the most oxidized and stable form of nitrogen in this environment. As this nutrient is the primary form of nitrogen used by plants, which convert nitrate into organic nitrogen to stimulate growth, excessive amounts can result in mass proliferations of phytoplankton and aquatic macrophytes. The main sources of pollution are soil fertilizers, human and animal sewage, and atmospheric deposition (BRITISH COLUMBIA, 1998).

5.5.1.2 Biological Variable (BV)

Analysis of physico-chemical factors contributes to understanding the spatial distribution of phytoplankton organisms, the basis of the trophic web in aquatic environments, and is of fundamental importance for qualitative and quantitative assessment of the level of degradation suffered by a river basin.

The result obtained for the biological variable was 0.60 (Table 5.26); this value was also observed in sub-basin 4 and represents average diversity. The artificial enrichment of nutrient salts resulting from the indiscriminate discharge of domestic effluents "in natura" into this environment leads to changes in the specific composition of the algal populations, with consequent alterations in their structure, diversity, biomass, metabolism and physiology (TUNDISI, 1986).

TABLE 5.26 - **Values of the biological variable per collection (density - Shanonn diversity index), value of the indicator per collection and average of the indicators for sub-basin 5.**

BIOLOGICAL VARIABLE - PHYTOPLANKTON		
COLLECTIONS	**DIVERSITY INDEX SHANNON (nits.cel-1)**	**VALUE OF INDICATOR**
1	2,43	1
2	1,97	0,7
3	1,55	0,7

4	0,59	0
INDICATOR AVERAGE		**0,60**

Estuaries are among the most fertile regions in the world and their productivity is an important means of life for the riverside populations who live on their banks and derive most of their sustenance from them. However, despite the importance that these ecosystems play for human populations and for the various species of organisms that depend on estuaries, they have been the targets of numerous anthropogenic impacts, acting as receptors for domestic and industrial effluents that have greatly altered their environmental conditions.

5.5.1.3 Environmental Preservation Index (IP)

In this sub-basin, a large anthropized area can be seen (6.38 km^2) to the detriment of vegetated and flooded areas; compared to the other sub-basins, it was the second worst value. Disorderly and irregular urban occupation has created problems, causing environmental systems to become unbalanced, burdening public authorities and posing risks to the population (Table 5.27).

TABLE 5.27 - **Environmental Preservation Index (IP) values for sub-basin 5.**

ENVIRONMENTAL PRESERVATION INDEX - IP	
LAND USE CLASSES	**VALUES**
Drainage Density (Dd)	2.62 Km/ Km^2
Vegetated Area (VA)	2.84 Km^2
Soaked Areas (AE)	2.97 Km^2
Anthropized Areas (AA)	6.38 Km^2
Environmental Preservation Index (IP)	**0.34** Km^2

Among them are those related to the physical environment and indiscriminate anthropic activities, such as the occupation of areas of floodplains, areas subject to flooding, areas with steep slopes and areas of soil susceptible to erosion processes (VALÉRIO FILHO *et al.*, 2002).

Flooding in urban areas represents a serious problem for most Brazilian municipalities, especially when it affects densely occupied areas, when it causes considerable and often irreparable damage, with loss of human life (CARVALHO *et al.,* 2001).

The Sà Viana neighborhood, located near the campus of the Federal University of Maranhão (UFMA), occupied an irregular area where there was mangrove swamp. The lack of maintenance of the dam, combined with the rainy season, caused flooding in the neighborhood.

Flood control in urbanized basins involves a set of structural and non-structural measures that range from urban and rural land use planning to human relocation and living with the problem.

From the graphic representation (Figure 5.5) it can be seen that despite being a region where occupation is recent, it has taken place at an accelerated rate, causing the degradation of natural resources, especially in areas close to rivers.

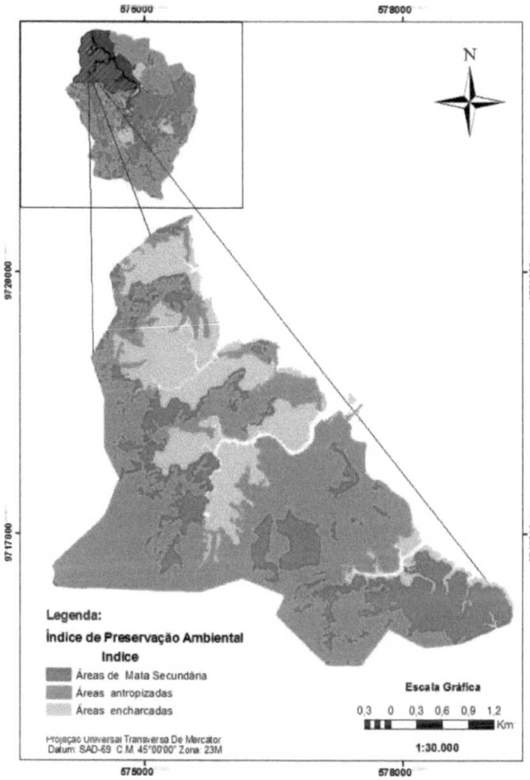

FIGURE 5.5 - Graphical representation of the Environmental Preservation Index for sub-basin 5.

The ED found for sub basin 5 was 0.2, showing that this sub-basin has poor sustainability O'Connor (1999) emphasizes the importance of environmental quality as a precondition for the well-being of human society and its long-term (economic) sustainability.

5.5.2 Socio-economic Dimension (SD)

Recent years have seen a series of changes due to technological development, disorderly population growth, economic crises, the formation of degraded peripheries as a result of social inequality and the phenomenon of metropolization due to the strong pull exerted by large urban centers.

This phenomenon has generated, and continues to generate, economic, social and environmental problems. The worsening of these problems is a direct result of the lack of adequate policies to tackle the deterioration in quality of life, especially in the main centers of economic activity.

The bridge over the Anil River, which paved the way for the conquest of the coastal strip by the middle and upper classes, has its opposite in the Bacanga River dam, which provided access to the port of Itaqui and made areas available for popular housing developments, of which the Anjo da Guarda neighborhood, with self-built housing, was the first and most important. This was followed by numerous irregular occupations in the area, generically known in São Luis as villas and identified by geographical references (Vila Embratel) or political names (Vila Mauro Fecury), attracted by the indirect jobs produced by the port and the construction of the Federal University campus (BURNETT, 2002).

5.5.2.1 Income Indicator (IR)

The distribution of the number of families according to per capita family income classes is an important indicator for supporting policies aimed at reducing poverty, inequality and regional differences. Therefore, the quantification of the population whose income is below a certain level is an essential aspect of sustainable development, insofar as the eradication of poverty and the reduction of inequalities are national and universal objectives (IBGE, 2009).

The IR found for this sub-basin was 0.02 (Table 5.28), the lowest value found for the Bacanga river basin; only the Vila Isabel neighborhood had an income between one and three salaries.

TABLE 5.28 - **Income indicator values for sub-basin 5.**

INCOME INDICATOR - IR	
NEIGHBORHOODS	**INDICATOR**
Sà Viana	0
Vila Isabel	0,25
Vila Dom Luis	0
Vila Bacanga	0
Sâo Raimundo	0
Gancharia	0
Guardian Angel	0

Vila Nova	0
Jambeiro	0
Alto da Esperança	0
TOTAL	0,25
INDICATOR AVERAGE	**0,02**

According to IPEA, the monthly per capita income required for an individual to be part of the country's richest 10% is R$ 571, meaning that a family of four with a family income of R$ 2,284 belongs to the richest 10% of the population. Of the heads of household, 83% are in a situation of extreme poverty in the country and either work for themselves or are unregistered workers.

Of the sub-basins studied, this was the one with the worst income indicator: of the ten neighborhoods, only Vila Isabel had a value of 0.25 (between 1 and 3 minimum wages). This shows that this population is extremely poor.

5.5.2.2 Education Indicator (EI)

A lot of research is focused on the problems of social reproduction. The behaviors of falling behind in school, dropping out of school, young people who don't work and don't study and don't look for work, are reflections of poverty. These behaviors are affected by the social composition of the neighborhoods.

Primary education is predominant in this sub-basin (Table 5.29). This reality is worrying, as it identifies a low level of education,

As a result, they have a very limited level of information and don't understand the minimum necessary to live as citizens and enjoy their rights and duties.

TABLE 5.29 - **Education indicator values for sub-basin 5.**

EDUCATION INDICATOR - IE	
NEIGHBORHOODS	**INDICATOR**
Sà Viana	0,3
Vila Isabel	0,3
Vila Dom Luis	0,3
Vila Bacanga	0,3
Sâo Raimundo	0,3

Gancharia	0,3
Guardian Angel	0,3
Vila Nova	0,3
Jambeiro	0,3
Alto da Esperança	0,3
TOTAL	3
INDICATOR AVERAGE	**0,3**

5.5.2.3 Sanitation Indicator (SI)

Water supply is a very serious problem in this sub-basin, because even though the vast majority of people are connected to the network, the supply is insufficient because the population is growing every day and occupying places that are difficult to channel, often creating clandestine wells or using water from neighbors who have a supply (Table 5.30). Many residents don't have water meters connected to their homes, which facilitates wastage, since the service provider can't assess the amount of water used in these places.

One of the biggest problems facing urban centers in Brazil is the lack of sewage systems. According to the IBGE (2009), 52% of municipalities collect household sewage, which is almost always disposed of without any treatment, mainly in rivers.

Of the ten neighborhoods in this sub-basin, only the Anjo da Guarda neighborhood had a higher score for water supply (1) and sewage network (0.15). Emergency measures should be taken by the competent bodies, as the health of the population and the quality of the water are related to these services.

TABLE 5.30 - **Sanitation indicator values for sub-basin 5.**

SANITATION INDICATOR		**- IS**
NEIGHBORHOODS	**WATER**	**SEWAGE**
Sà Viana	0,7	0
Vila Isabel	0,6	0
Vila Dom Luis	0,8	0
Vila Bacanga	0,6	0
Sâo Raimundo	0,6	0

Gancharia	0,7	0
Guardian Angel	1	0,15
Vila Nova	0,75	0
Jambeiro	0,8	0
Alto da Esperança	0,5	0
TOTAL	7,05	0,15
INDICATOR AVERAGE	**0,70**	**0,01**

This sub-basin also had a low DS value (0.20), since this population does not have the necessary essential services, such as water supply, which is precarious throughout the sub-basin; there is no sewage system and although there is solid waste collection, residents dispose of their waste in the open, reflecting the lack of environmental awareness.

The lack of implementation of policies aimed at the issue of popular housing has led to the outbreak of a housing deficit. The majority of the territories occupied by the low-income classes have culminated in sub-standard occupations, putting most of this population at risk of landslides, floods, contamination by pollutants, among others.

5.6 Sub-basin analysis and ISAtotal

The application of the ISA to the five sub-basins studied, in addition to allowing specific conclusions to be drawn about the sustainability of each sub-basin, also makes it possible to establish a comparative analysis between the sub-basins. This analysis can achieve various objectives, such as establishing criteria for ranking the sub-basins with the greatest or least degree of conflict, identifying the sub-basins in need of priority action and analyzing sustainability trends for the sub-basins.

Sub-basin 5 had the lowest ISA (0.29) compared to the other sub-basins. This low value can be attributed to the indicators of income, schooling and sanitation. The data obtained on these indicators shows that low levels of income and schooling force people to occupy risky areas without infrastructure, such as hillsides and riverbanks.

The best ISA result was obtained in sub-basin 1 (0.43), which is due to the better socio-economic conditions of the population, since the best results for DS were found in this sub-basin. However, as for ED, this dimension showed the lowest result, especially for IP (0.03), the anthropized area is 5.18, to the detriment of the sum of the area of secondary forest and flooded, showing that the environmental quality of the sub-basin is very compromised.

A value close to that of sub-basin 1 was found in sub-basin 2, with an ISA of 0.42, its dimensions also having the same value. The highest phosphate (3.37) and ammonia (1.92) values were found in this sub-basin. Levels inconsistent with CONAMA Resolution 357/05 were also found for dissolved oxygen.

Particular attention should be paid to sub-basin 3, as it is a UC. The Recanto Verde, Vila Itamar and Vila Esperança urban agglomerations located near the springs that feed the Batatâ reservoir are compromising the environmental quality of local water bodies through domestic sewage.

Cavalcante (2005) carried out work in sub-basin 3 and observed the presence of thermotolerant coliforms in 58.3% of the data sampled. The physical-chemical variables pH, DO, total phosphorus and total iron showed alterations that are not in line with the drinking water standard.

Sub-basin 4 showed the same result as sub-basin 3 (0.36). The socio-economic dimension (SD) obtained a low result (0.25), showing that more incentives should come from public bodies to subsidize this needy population by offering sanitation services, income and education.

In the process of urbanization of the Bacanga river basin, there are two situations: on the right bank there are areas of consolidated urbanization, with densities close to saturation, becoming less dense as you get closer to the Sacavém neighbourhood, which is nevertheless expanding rapidly; and on the left bank there are rarer areas, with the exception of the Anjo da Guarda and Vila Embratel neighbourhoods, which are already more consolidated. (MMT, 2007)

The graphic representation (Figure 5.6) of the Bacanga river basin shows that both the right bank (older occupation) and the left bank (more recent occupation) are densely populated.

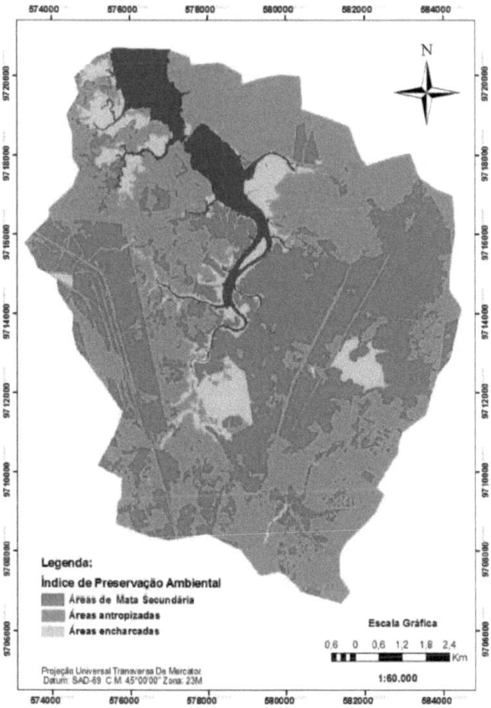

FIGURE 5.6 - **Graphical representation of the Environmental Preservation index (IP) for the Bacanga river basin.**

It is essential to explain that the population expansion of sub-basins 3, 4 and 5 is more recent and accelerated, with levels of degradation close to those found in sub-basins 1 and 2, which have older population agglomerations.

After a thorough analysis of the values for each sub-basin, it can be seen that the ISA found for the Bacanga river basin as a whole was 0.37 (Table 5.31), thus revealing weak sustainability, which is that which assumes the unlimited possibility of substitutes between the different forms of capital

TABLE 5.31 - **Values for the Ecological Dimension (ED), Socioeconomic Dimension (SD) and Water Use Sustainability Indices by sub-basin (ISAsub) for the Bacanga river basin.**

SUB-BASINS	DE	DS	ISAsub
SUB1	0,24	0,63	0,43
SUB2	0,42	0,42	0,42
SUB3	0,50	0,23	0,36

SUB4	0,47	0,25	0,36
SUB5	0,38	0,20	0,29
ISAtotal	0,40	0,34	0,37

According to BIDONE *et al.* (2004) the total capital of an economy is described by the sum of different forms of capital, namely: man-made capital (goods and services generated by man); human capital (measured by the average education level of the population); natural capital (natural resources) and social and moral capital. Strong sustainability (in addition to the increase in man-made capital and human capital) assumes that capital should not be irretrievably replaced by other forms of capital.

Hence, a system is considered to be more or less sustainable to the extent that it is able to maintain or improve the quality of the environment, mitigate degradation and anthropogenic impact, reduce social inequality and provide inhabitants with basic living conditions, i.e. a healthy and safe built environment, as well as build political pacts that make it possible to face present and future challenges.

In this way, we can infer that the environment contains our possibilities for development and improving our quality of life, as well as that of future generations. Because every productive system depends, directly or indirectly, on natural resources.

According to the National Environmental Policy, a polluted environment is one that presents a degradation of environmental quality resulting from activities that directly or indirectly harm the health, safety and well-being of the population; create adverse conditions for social and economic activities; adversely affect the biota; affect the aesthetic or sanitary conditions of the environment; release materials or energies in disagreement with established environmental standards. All these conditions are verified in the Bacanga river basin, so it can be classified as a polluted basin.

In this region, the discharge of domestic effluents, garbage disposal, landfills, mangrove cutting, as well as intense fishing activity, are the main factors of anthropogenic pressure which, to varying degrees of magnitude, contribute to the degradation and de-characterization of this important ecosystem.

Despite the degree of pollution found in the basin, Martins (2008), working with fishermen, considered the basin to be important from a social and economic point of view, as it promotes the livelihoods of the people living in the area, despite the reduced environmental quality.

5.7 Mitigation measures

In the end, it was considered that the delays in local municipal action, with the absence of programs and financing aimed at the low-income population, as well as the lack of convergence of efforts in metropolitan terms by the different municipal spheres (including the state) to consolidate a Metropolitan Housing Policy, led to a chaotic and growing scenario of disorderly occupations and invasions of environmentally unsuitable areas.

In the basin under study, there are two programs being carried out: the Bacanga Basin Environmental Recovery and Quality of Life Improvement Program, which is structured along urban, socio-economic, environmental and institutional lines of action, based on the concepts of sustainable development and human development, developed by the Sâo Luis City Council and the World Bank (BIRD). The other program is the PAC do Bacanga, a federal government program whose executing unit is the Sâo Luis City Hall. Its priorities are water supply and sewage treatment in the Itaqui-Bacanga area (the left bank of the basin), as well as actions aimed at the socio-economic aspect of the population.

As all the sub-basins and consequently the Bacanga river basin obtained a value below 0.5, it can be seen that the mitigating measures are for the Bacanga basin. Among the mitigating measures, we can mention:

- Local economic development through the generation of work and income by training people outside the formal labor market, for example, as guides in the UC;
- Digital inclusion with the implementation of telecenters;
- Articulation of the social protection network, so that families can be registered at municipal and state level;
- Expansion and improvement of water supply services, sewage collection and treatment, as well as urban drainage to avoid contamination of water bodies;
- Recovery of the Batatâ water treatment plant (WTP), as it is currently operating at only 43% of its total capacity, and is responsible for supplying many neighborhoods on the right bank of the Bacanga river basin;
- Recovery of the Bacanga river dam to normalize water flow and prevent flooding;
- Relocation of residents from areas at risk, such as bayous and flooded areas;
- Respect for the boundaries of the Bacanga State Park;
- Adoption of solutions that contribute to the recovery and preservation of protected areas and areas of environmental interest.

6 FINAL CONSIDERATIONS

Population growth and the resulting increase in the waterproofed area in the Bacanga river basin, the silting up of its riverbeds, the pollution of water bodies and deficiencies in urban drainage planning form a picture of the main problems afflicting many Brazilian municipalities.

The history of occupation of the area under study reveals progressive and disorderly growth, as well as the absence of an efficient policy on the part of the public authorities. Water supply, even with inadequate distribution, was the best service provided to the population, while the sewage system was the worst. Sanitation is an important component in improving the quality of life of the population and environmental conditions in the area under study.

In the current scenario and with regard to opportunities for entering the job market, work is considered to be one of the few opportunities for social mobility for low-income families. Education has been reaffirmed as the only instrument for economic and productive development in response to poverty.

Due to the nature of the problems, obstacles and possibilities inherent in urban management, he also adds three interrelated dimensions to this concept, highlighting its multidisciplinary nature: political-institutional management, political-social management and political-administrative management.

In this sense, it is clear that it is important to manage all the city's spaces, but with special attention to those that fall outside conventional urban planning standards, thus requiring more specific actions to mediate social, economic, environmental and urban conflicts.

Environmental perception is an important field of study for better understanding the interrelationships between man and his environment. Manifestations of dissatisfaction with the city's built environment are constant due to the poor quality of urban spaces, especially in peripheral areas.

The basis for studies and research into environmental education is based on the understanding that human beings' experience of their environment is instructed by perception. People, both in urban and non-urbanized environments, recognize environmental conditions through their perceptual processes. They begin to process in their minds what is perceived through their sensations and progressively acquire an understanding of the environment that surrounds them, in other words, an understanding of the anthropization of nature, the occupation of a natural territory by human beings, transforming it into a cultural landscape.

The monitoring methodology developed based on the ISA allows corrective or mitigating action to be taken when negative environmental impacts are detected. The research has generated an important database on the sustainability of the Bacanga river basin, which can be fed with spatial and temporal information and will make it possible to use more indicators.

The sustainable development of water use in the Bacanga river basin is a process under construction, the quantification and qualification of which has been expressed through two dimensions. The concept of sustainability using the ISA was developed with the aim of classifying the sub-basins, the quality of the water in the Bacanga river basin, environmental quality and sources of pollution.

The main result of applying this methodology was a significant increase in the demand for and alteration of surface water, due to the urban needs of the riverside populations along the Bacanga River. There is a consensus that the quality of water is being degraded by anthropogenic activities, but the degree of impact of these activities can be reduced by implementing management practices and conserving natural resources, as well as applying and complying with existing legislation.

It should be noted that this analysis is capable of indicating the possible causes of unsustainable water use, whether in sub-basins with good socio-economic indicators or in relation to their ecological aspects. For the basin under study, the following considerations are made regarding the low value obtained by the ISA:

- Deficiency in the education system;
- Low level of investment in basic services (water supply and sanitation);
- Environmental impacts resulting from the inadequate disposal of urban effluents and domestic solid waste, associated with income concentration and susceptibility to urban pollution.

There should be greater social investment in areas where the development of economic activities is incorporated into the dynamics of more intensive land occupation.

Finally, it can be said that the great advantage of the ISA is that it can be used for strategic evaluation in public policies and for environmental management groups. It supports the management of water resources in river basins, allowing the establishment of priority areas for monitoring water quality, as well as identifying the variables that should be prioritized in these studies, facilitating intervention in the surrounding area.

7 conclusions

- The Bacanga river basin is polluted due to the impairment of its sub-basins, caused by the lack of planning for urban expansion and consequently the absence of a sewage system, lack of treatment of domestic effluents, inefficiency in water distribution, exposure of the soil due to deforestation of vegetation, lack of supervision in the UC, low-income population, no schooling, unskilled labor, no prospect of better employment living in risky places;
- In the sub-basins, low-income and low-education populations occupy peripheral areas, often at risk of death;
- Although sub-basin 1 had the best ISA, the ecological dimension was very compromised due to its large deforested area, i.e. low environmental preservation index;
- Sub-basin 2 showed results incompatible with those of CONAMA Resolution 357/05 for the physico-chemical variables phosphate, ammonia and DO, with the worst values observed in the entire basin, thus demonstrating a high degree of environmental pollution;
- The lack of monitoring in sub-basin 3 has contributed to the fact that the boundaries of the Bacanga State Park have shrunk over time;
- The construction of the Bacanga river dam led to the unplanned expansion of the Itaqui-Bacanga area;
- The expansion of a deprived population, together with a large area of exposed soil due to deforestation for the construction of houses and the favoring of flooding, together with the inefficiency of sewage collection, treatment and water supply services, contributed to sub-basins 4 and 5 having the worst ISA;
- The ISA methodology proved to be efficient, but it needs to be improved by giving greater weight to the physico-chemical and biological variables, including indicators such as BOD, COD, heavy metals, microbiological water quality; in the socio-economic sphere, health indicators, social projects, political participation could also be included; these indicators would give a more complete result on the socio-economic environmental situation;

REFERENCES

ALCÂNTARA, E. H. **Aplicaçâo do sensoriamento remoto e da geoestatistica no estudo de variâveis fisico-quimicas e biológicas do estuàrio do rio Anil. Sao Luis, Maranhao - Brazil**. Sâo Luis, 2004. 53p. Monograph (Bachelor's Degree in Aquatic Sciences), Federal University of Maranhâo.

ALCÂNTARA, E. H. AMORIM, A. J. de.; Morphometric analysis of a coastal hydrographic basin: a case study. **Caminhos da geografia- Online journal**. 7077, feb. 2005. Available at <www.ig.uiu.br/caminhos_da_geografia.html> Accessed on November 13, 2009.

ANTONELI, V; THOMAZ, E.L. Characterization of the physical environment of the Arroio Boa Vista basin, Guamiranga-PR. **Rev. Caminhos da Geografia,** Uberlândia, v.8, n.21, p46-58, jun. 2007.

AZEVEDO, J. de. **Tool for the analysis of socio-economic and environmental data for the definition of public policies**, Case Study, Environmental Basin of the Imboassu River, Municipality of Sâo Gonçalo/RJ, 2006. 200p. Thesis (Doctorate), Fluminense Federal University.

BAUMGARTEN, M. G. Z.; ROCHA, J. M. B.; NIENCHESKI, L. F. H. **Manual of analysis in chemical oceanography.** Rio Grande do Sul: Ed. FURG, 1996. 132p.

BENDO, A., *et al.* 2000. Dissolved nutrients in Lagoa da Conceição and the central region of Florianópolis, SC. Itajai, **Proceedings of the 13th** National Oceanography Week, 2000**,** p. 249-251.

BENETTI, L. B. **Evaluation of the sustainable development index of the municipality of Lajes (SC) using the sustainability panel method.** Santa Catarina**,** 2006. 198p Thesis (Doctorate), Federal University of Santa Catarina.

BIDONE, E. D.; MORALES, P. R. D. **Desenvolvimento Sustentàvel e Engenharia (Enfoque Operacional)** Rio de Janeiro: Fundaçâo Ricardo Franco, 2004. 260p.

BIDONE, E. D.; CASTILHOS, Z. C.; AZEVEDO, J. de. Socio-economic assessment of environmental impacts in Pressure - Conditioning - Impact - Response (PCIR) structures. ROMEIRO, A. R. (Org). In. ---Evaluation **and accounting of environmental impacts** Sâo Paulo, SP; Imprensa Oficial do Estado de São Paulo, 2004, p. 183 -195.

BITTON, C. C.; **Statistical tools and simulation model of the environmental processes of the Piratininga and Guarapina Coastal Lagoons - RJ - Brazil.** Thesis (Doctorate), Fluminense Federal University, 2000. 233p.

BRAGA, T. M.; FREITAS, A. P. G. de; DUARTE, G. de S.; SOUSA, J. C. **Nova Economia_Belo**

Horizonte. 14(3)_11- 33_September - December 2004.

BRAGA, E. S. **Dissolved nutrients and primary production of phytoplankton in two estuarine systems of the State of São Paulo**. 1995, 120p. Thesis (PhD), Oceanographic Institute of the University of Sâo Paulo, Sâo Paulo.

BRANCO, S. M. **Hidrobiologia aplicada à engenharia sanitària**. 3 ed. Sâo Paulo, CETESB/ASCESTEB, 1986. 640P

BRAZIL. **CONAMA Resolution No. 357**, of March 17, 2005. Provides for the classification of bodies of water and environmental guidelines for their classification, as well as establishing the conditions and standards for the discharge of effluents, and makes other provisions, 2005;

BRITISH COLUMBIA (Canada). **Guidelines for interpreting water quality data**: version 1.0, 1998. Available at: <http://srmwww.gov.bc.ca/rise/pubs/aquatic/interp/interp-01.htm> Accessed on 22 Oct. 2009.

BURNETT, C. F. L.; **Além do rio Anil, urbanização e Desenvolvimento Sustentàvel: estudos sobre a sustentabilidade dos tipos de urbanização na cidade de Sâo Luis do Maranhâo.** 2002, 200p Dissertaçâo (Mestrado) Universidade Federal de Pernambuco.

BERGESCH, M.; ODEBRECHT, C. Analysis of phytoplankton, protozooplankton and some abiotic factors in the Patos Lagoon estuary. **Atlàntica**, 19: 31-50.

CAMARGO MORA, M. G. Calidad de vida y capacidades humanas. **Revista geogràfica venezoelana** - 040 (2) 1012-1617; 247-258, biannual, 2003.

CARMOUSE, J. P. **The metabolism of aquatic ecosystems. Theoretical foundations, study methods and chemical analysis**. Sâo Paulo: Editora Edgard Blücher / FAPESP, 1994. 255 p.

CARVALHO, P. F. de; BRAGA, R. (eds.) **Perspectivas** de **Gestâo Ambiental em Cidades Médias**. Rio Claro: LPM-UNESP, 2001, 95-109.

CAVALCANTE, A. de J. B. D. **Uso e ocupação da micro-bacia do Batatâ, no Parque Estadual de Bacanga (Sâo Luis, MA) e seus reflexos na degração dos recursos hydricos**. Sâo Luis, 2005, 75 p. Monograph (Graduation in Aquatic Sciences) Federal University of Maranhão;

CHESTER, R. **Marine Geochemistry**. Unwin Hyman, London: 346-421p, 1990.

COELHO, C. J. C. **Aspectos da disponibilidade e dos usos da água na bacia do rio Bacanga/Ilha do Maranhâo) Ilha de Sâo Luis - MA.** Sâo Luis, 2006, 125p. /Monograph (Bachelor in Aquatic Sciences) Federal University of Maranhâo;

COSTA, P. R. **Distribution of the phytoplankton community in relation to physico-chemical**

factors in the Bacanga river estuary, Sâo Luis/MA. Sâo Luis, 2008, 50 p. Monograph (Graduation in Aquatic Sciences) Federal University of Maranhão;

COUTO, O. F. V. **Generation of an environmental sustainability index for watersheds in urban areas using integrated geoprocessing techniques**. Curitiba, 2007. 172p. Dissertation (Master's Degree), Federal University of Rio Grande do Sul;

CHRISTOFOLETTI. A. THE analysis of drainage density and its geomorphological implications. **Geografia.** v.36. n.8. 1979, p.23-42.

DIAS, G.F. **Atividades Interdisciplinares de Educaçâo Ambiental: Manual do Professor.** Sâo Paulo: Global/Gaia, 2004. Available at:

<http://www.ciesin.columbia.edu> Accessed on: 01 Mar 2010.

ESCHRIQUE, S.A.; **Hydrogeochemistry of phosphorus in the Jaguaribe estuary (CE).** 2007. 100p Dissertation (Master) Federal University of Cearà,

ESI, **Environmental Sustainability Index** - an initiative of the global leaders of tomorrow environmental task force. (In collaboration with: Yale Center for Environmental Law and Policy Yale University and Center for International Earth Science Information Network Columbia University) 2002.

ESKINAZI-LEÇA, E., FRANCA, L. M. B., MACÊDO, S. J. Ecological study of the Itamaracà area (Pernambuco-Brazil). XXIV standing stock of phytoplankton in the Botafogo River estuary, January to December 1975. **Trabalhos Oceanogràficos da Universidade Federal de Pernambuco**, Recife, v.18, p.153-192, 1984.

ESKINAZI-LEÇA, E.; KOENING, M. L.; SILVA-CUNHA, M. da G. G. Marine microalgae from Pernambuco: The vegetation you never see. In: ARAÙJO, E. de L.; MOURA, A do N.; SAMPAIO, E. V. de S. B.; GESTINARI, L. M. de S.; CARNEIRO, J. de M. T. **Biodiversidade, conservação e uso sustentâvel da flora do Brasil**. Recife: SBB/EDUFRPE, 2002. p. 186-189.

ESTEVES, F. de A. **Fundamentos de Limnologia.** 2. Ed. Rio de Janeiro: Interciência, 1998. 601p.

FAY, E. F. Sustainable water use index (ISA-AGUA) in the sub-medium Sâo Francisco region. Jaguariûna: **EMBRAPA**, Meio Ambiente, 2006. 157p,

FERNANDES, D. R. **A contribution on the construction of indicators and their importance for business management.** Revista da FAE, Curitiba, V.7, N.1, P.1-18, jan./jun., 2004.

FIDALGO, E. C. C. **Criteria for the analysis of environmental methods and indicators used in the diagnostic stage of environmental planning.** Campinas, 2003. 276f. Thesis (Doctorate in

Agricultural Engineering) - Faculty of Agricultural Engineering, State University of Campinas.

FRÓES, J. F. **Impactos sócio-urbanos no alto curso do Rio das Bicas: contextualizando o bairro do Coroadinho**. Sâo Luis, 2004, 189p. Monograph (Graduation in Geography) - Federal University of Maranhão.

HERCULANO, S. Quality of life and its indicators. **Ambiente & Sociedade**, v.1, n° 2,1. Sem. 1998.

HORNE, A. J.; GOLDMAN, C. R. **Limnology.** 2.ed. Singapore: McGraw-Hill, 1994.

IBGE. Brazilian Institute of Geography and Statistics. **Cities**. 2008. Available at: <http://www.gov.br/cidadesat/topwindow.htm?1>. Accessed on: May 13, 2009.

IBGE. Brazilian Institute of Geography and Statistics. **Cities**. 2009. Available at: <http://www.gov.br/cidadesat/topwindow.htm?1>. Accessed on: May 13, 2009.

IBGE. Brazilian Institute of Geography and Statistics. **Census data.** 2002 - Sâo Luis/MA.

IPEA - Institute of Applied Economic Research, <http://www.ipea.gov.br/pub/td/td.html> Accessed on: 15 Jan. 2010.

ISAIAS, F. B. **The sustainability of water: a proposal for a sustainability index for river basins**. Brasilia/DF, 2008. 168p Thesis (Doctorate) University of Brasilia.

ISLA, M. **A review of the urban indicators experience and a proposal to overcome current situation. The application to the municipalities of the Barcelona province**. (Paper to World Congress of Environmental and Resource Economists) Isola de San Giorgio, Venice, Italy: June 25-27, 1998.

KREBS, C. J. **Ecological methodology**. New York, Harper Collins, 1989, 645p.

LABOHIDRO - Hydrobiology Laboratory of the Federal University of Maranhão. Technical Report: **Environmental Diagnosis of the Bacanga River Estuary, São Luis Island - MA**. São Luis, 1998.

LANA, C. E.; ALVES, J. M. de P.; CASTRO, P. de T. A. Morphometric analysis of the Rio do Tanque basin, MG - Brazil. **Rem: Rev. Esc. Minas**, Ouro Preto, v. 54, n. 2, June 2001.

MACEDO, S. J. COSTA, K. M. P. Ecological studies of the Itamaracà region, Pernambuco - Brazil. Hydrological conditions of the Botafogo River estuary. **Ciência Cultura** 30: 346 - 368, 1978.

MACÊDO, L.A.A. **Qualidade ambiental dos rios da Ilha de Sao Luis**. Sâo Luis; UFMA, 2003. 76p.

MAGALHÂES JÙNIOR, A. P., CORDEIRO NETTO, O. M., NASCIMENTO, N. O. Indicators as potential tools for water management in Brazil's current legal and institutional context - results from a panel of experts. **Brazilian Journal of Water Resources (RBRH).** Vol. 08, n. 04, Oct/Dec, 49 - 67p. 2003.

MARGALEF, R. **Temporal succession and spatial heterogeneity in phytoplankton**. In: BUZZATI--TRAVERSO, A.A. (Ed.). Perspectives in marine biology. Berkely: California University Press, p.323-349. 1958.

MARGALEF, R. 1983. **Limnology**. Barcelona, Omega S. A.1010p.

MARINS, R V.; ABREU, I. M.; AGUIAR, J. E. Distribution of carbon and phosphorus in estuarine sediments of the Jaguaribe River - Cearà. In: **Livro de resumos - A Quimica torna sua vida melhor 26a Reuniâo Anual da SBQ,** Poços de Caldas. Sociedade Brasileira de Quimica, Sâo Paulo: 2003, 1p.

MARTINS, A. L. P. **Assessment of the environmental quality of the Bacanga watershed (Sâo Luis - MA) based on physico-chemical, biological and population variables: subsidies for sustainable management.** Sâo Luis, 2008, 113p. Dissertation (Master's Degree in Ecosystem Sustainability), Federal University of Maranhão.

MMT Planning and Consulting. Environmental consultancy report. **Program to recover and improve the quality of life in Bacanga.** Sâo Luis, 2007.83p.

mo.htm# FOOTNOTE> Accessed: Sept. 28, 2009.

MORAES, D. S. de L. & JORDÂO, B. Q. Degradation of water resources and its effects on human health. **Rev. Saùde Pùblica** 36(3):370-374, 2002.

MOTTA, R. S. da. Environmental indicators in Brazil: ecological, efficiency and distributive aspects. **Text for discussion n° 403**. ISSN 1415-4765 Rio de Janeiro - RJ - IPEA, 1996.

MOTA, A. da S. **Coroadinho: History of land occupation from the 1940s to the present day.** Sâo Luis, 1995, 149p. Monograph (Graduation in History) Federal University of Maranhão.

O'CONNOR, M. **Natural capital**. Policy Research Brief Series, n. 3, Cambridge Research for the Environment, 1999, 22 p.

OLIVEIRA, C. C. C. de. **The pattern of urban occupation and its influence on the riparian vegetation of the Bacanga River.** Sâo Luis, 2008, 50 p. (Monograph in Aquatic Sciences) Federal University of Maranhão.

PAULA FILHO, F. J. **Distribution and geochemical participation of phosphorus in estuarine sediments of the Rio Jaguaribe/CE**. Fortaleza, 2004, 88p. Dissertation (Master's Degree in Tropical Marine Sciences). Federal University of Ceará.

PEREIRA, L. S. **Mobilização social para a gestão ambiental: construção de uma metodologia a partir da discutâo dos problemas da bacia hidrogràfica.** Sâo Luis, 2003. 79 p. Monograph (Specialization in Water Resources and Environmental Management) Universidade Estadual do Maranhâo.

PORTILHO, F. **"Green"** consumption**, ecological democracy and citizenship: possibilities for dialogue?** Available at:< www.rubedo.psc.br> Accessed: Oct. 25, 2009.

PRETECEILLE, E., RIBEIRO, L. C. de Q. Trends in social segregation in global and unequal metropolises: Paris and Rio de Janeiro in the 80s. **Revista Brasileira de Ciências Sociais**, v. 14, n. 40, p. 143-162, Anpocs, jun. 1999.

RAMIERI, E.; COGO, V. Indicators of Sustainable Development for The City and The Lagoon of Venice. **(Paper to World Congress of Environmental and Resource Economists)** Isola de San Giorgio, Venice, Italy: June 25-27, 1998.

RHAMA Environmental Consulting. Final Report. **Hydrodynamic and water quality modeling of the estuarine mouth of Sâo Luis/MA**. 2008. 141p.

RIBEIRO JÛNIOR, J. **Formaçâo do espaço urbano de Sâo Luis: 1612 - 1991**. Sâo Luis: Ediçoes Func, 1999.

ROSS, J. L. S. Empirical analysis of the fragility of natural and anthropized environments. In: **Revista do Departamento de geografia**, n° 8, FFLCH/USP, 1994, p. 63 - 74.

SANTOS, T. G.; BEZERRA - JÛNIOR, J. L.; COSTA, K. M. DA.; FEITOSA, F. A. do N. Dynamics of phytoplankton biomass and environmental variables in a tropical estuary (Pina Basin, Recife/ PE) **Revista Brasileira de Engenhaia de Pesca** 4 (1), jan 2009.

SILVA, J. P. R. P. da. **Environmental risks and disasters caused by flooding in urban areas**: a study of the landscape from a systemic viewpoint. 2009, 75 p. Monograph (Bachelor's Degree in Geography) - Paulista State University, Rio Claro.

TAYLOR, D. Using Sustainability Indicators to Implement Local Agenda 21. **(Paper to World Congress of Environmental and Resource Economists)** Isola de San Giorgio, Venice, Italy: June 25-27, 1998.

TUCCI, C.E.M. **Avaliaçâo e controle dos impactos ambientais decorrentes da urbanizaçao: apresentação do projeto.** In: Evaluation and control of urban drainage. Porto Alegre: Ed. UFRGS, 2000. p. 15-21;

TUNDISI, J. G. Estuarine plankton. **Contribuiçôes Avulsas [do] Instituto Oceanogràfico da Universidade de Sâo Paulo**, Sâo Paulo, n. 19, p. 1-22, 1970. (Biological Oceanography Series).

1970.

TUNDISI, J. G. Ecological studies of marine and lacustrine phytoplankton in Brazil. In: BICUDO, C. E. M.; TEIXEIRA, C.; TUNDISI, J. G. (Ed.) **Algae: the energy of tomorrow**. Sâo Paulo: [s.n.], 1986. p. 27-48.

TURNES, V. A. **Delos System: indicators for sustainable local development processes.** 2004. 237p. Thesis (Doctorate), Federal University of Santa Catarina.

VALÉRIO FILHO, M.; ALVES, M.; GARCIA, R.; FANTIN, M. Characterization of watersheds sealed by the urbanization process with the support of geotechnologies. **Anais XI SBSR,** Belo Horizonte. INPE, p.1977 - 1983, 2003.

VILLELA, S.M.; MATTOS, A. **Hidrologia aplicada.** Sâo Paulo: McGRAWHill, 1975 150p.

WCED - World Commission on Environment and Development. **Our Commun future. The Brundtland Report**. Oxford University Press, 1987. 43p.

MIX
Papier aus verantwortungsvollen Quellen
Paper from responsible sources
FSC® C105338

Printed by Books on Demand GmbH, Norderstedt / Germany